Economics of Fibre Markets

Economics of Fibre Markets

A Global View of the Interdependence Between Man-made Fibres, Wool and Cotton

by

C. A. TISDELL

Professor of Economics
Department of Economics
University of Newcastle, Australia

and

P. W. McDONALD

Graduate Research Assistant
Department of Economics
University of Newcastle, Australia

PERGAMON PRESS

OXFORD · NEW YORK · TORONTO · SYDNEY · PARIS · FRANKFURT

U.K.	Pergamon Press Ltd., Headington Hill Hall, Oxford OX3 0BW, England
U.S.A.	Pergamon Press Inc., Maxwell House, Fairview Park, Elmsford, New York 10523, U.S.A.
CANADA	Pergamon of Canada, Suite 104, 150 Consumers Road, Willowdale, Ontario M2 J1P9, Canada
AUSTRALIA	Pergamon Press (Aust.) Pty. Ltd., P.O. Box 544, Potts Point, N.S.W. 2011, Australia
FRANCE	Pergamon Press SARL, 24 rue des Ecoles, 75240 Paris, Cedex 05, France
FEDERAL REPUBLIC OF GERMANY	Pergamon Press GmbH, 6242 Kronberg-Taunus, Pferdstrasse 1, Federal Republic of Germany

First Edition 1979

British Library Cataloguing in Publication Data

Tisdell, Clement Allan
Economics of fibre markets.
1. Fibers - Marketing
I. Title II. McDonald, P W
380.1'45'677 HD9155.A2 78-40925

ISBN 0-08-022468-7

In order to make this volume available as economically and as rapidly as possible the authors' typescripts have been reproduced in their original forms. This method unfortunately has its typographical limitations but it is hoped that they in no way distract the reader.

Printed and bound at William Clowes & Sons Limited
Beccles and London

Contents

List of Tables

List of Figures

Preface

This Pergamon monograph, *Economics of Fibre Markets: Interdependence Between Man-Made Fibres, Wool and Cotton,* updates, revises and extends our earlier monograph prepared at The University of Newcastle for limited circulation. Our research into fluctuations in wool prices and the nature of competition faced by the wool industry brought home to us the need for a volume of this kind. We found it difficult to obtain published material giving an up-to-date and global view of the marketing of, and economic interdependence between, fibres. Because it was essential background for our research into the wool industry, it was necessary for us to collect and analyse this material. This monograph brings together the results of our research. It provides background information on fibre markets generally and advances and tests new hypotheses about fluctuations in the prices of natural fibres and the marketing and promotion of fibres.

Our approach is to consider wool in the context of fibre markets as a whole. Therefore, even though this monograph focuses on the wool industry, it provides background information on the man-made fibre industry and the cotton industry and introduces theories and results applicable to these industries.

Institutional and empirical data as well as theory are introduced in this exposition but in a way which should make the main results interesting and accessible to anyone involved in, or concerned about, the operation of fibre markets. *Economics of Fibre Markets* has not been written exclusively for fellow economists, even though there are some technical sections (Chapters 3 and 9 and Appendices I and II) which the general reader may prefer to skim through rather than read in detail.

The research for the monograph was supported in part by a grant from Wool Industry Research Funds distributed through the Australian Wool Corporation and we are thankful for this assistance. We have also been fortunate in having the excellent facilities of The University of Newcastle available to us. Without these support facilities it would have been impossible to complete this study. The desirability of a study along these lines was first suggested to the senior author by Mr R.B. McMillan of the Australian Wool Corporation and early moves in this direction were encouraged by Dr Lionel Ward and Mr Kevin Foley of the Corporation. Without this early support, it is doubtful whether this research would have been started by the authors. The authors were fortunate in having the co-operation of these and other Corporation members during the period of research.

During study leave in 1975, when the senior author was based at The University of York, he had useful and stimulating discussion with Mr K. Clarke, Professor Jack Lewis, Dr John McPhee, Mr P. Bell and Mr J. Brown and other members of the International Wool Secretariat in England and Mr E. Kölsch, Mr H. Haarer and other members of the International Wool Secretariat in Germany. He also benefited from discussions with Mr J.T. Wenham, General Secretary, The Textile Institute, Manchester, Mr Wiseman of the Shirley Institute, Manchester, and Mr David Morgan and other members of I.C.I.'s Fibre Division at Harrogate. He had an opportunity to exchange ideas with Professor Ashton, Professor Carpenter and Mr I.M. Sturgess of University of Newcastle-on-Tyne, Dr J. Knapton and Dr Townsend of the University of Leeds, Professor Bottomley and Mr G. Oxtoby of Bradford University, Mrs Linden Briscoe of Manchester University and members of the Institüt fur Weltwirtschaft, Kiel.

In general, we have had a great deal of co-operation from all associated with the textile industry. When the project began and at a time when we had little idea of the operations of the textile industry, Mr L. Jeffreys and other members of Courtaulds Australia Ltd., were extremely helpful in providing us with background information. We were also encouraged by the fact that so many Australian textile producers participated in our surveys even though they were suffering extreme economic difficulties.

Members of the Australian Bureau of Agricultural Economics, several Australian University Libraries, Department of Agriculture and other Australian institutions provided us with a considerable amount of background material.

Our colleague, Mr Gary Keating, participated in some of our earlier research and we would especially like to thank him for his statistical assistance with our survey of Australian textile manufacturers.

We also extend our thanks to the Printing Section of The University of Newcastle for their assistance with photographic reductions.

Mrs Maureen Strazzari typed the whole manuscript cheerfully and efficiently. We thank her for the care she has taken in doing this.

Clem Tisdell,
Paul McDonald,
Department of Economics,
The University of Newcastle,
July 1978.

CHAPTER 1

New Fibres and the Patterns of Fibre Production and Consumption

1.1 INTRODUCTION

In order to understand the fundamental forces which have affected the marketing of natural fibres, especially wool, it is necessary to study the fibre market as a whole and in particular to take account of the development of man-made fibres. In considering some of the more important marketing problems encountered by natural fibres, mainly wool and cotton, we shall explore ways in which these are interconnected with the nature and structure of the man-made fibre industry. The principal problem discussed in this book, the presence of large and possibly growing fluctuations in the price of natural fibres, can only be adequately comprehended if the nature of man-made fibre production and marketing is specifically considered. This and the next Chapter provide a general world-wide background on the development, structure and nature of competition in the textile fibre industry.

Tremendous changes have occurred in the world fibre and textile industries in this century as the result of the development of man-made fibres. We are now reaching the stage where almost half of the textile fibres used in the world are man-made. The speed of market penetration by these fibres has been particularly rapid in the last 17 years. During this period, as indicated in Table 1.1, the market share held by man-made fibres doubled.

The growth of man-made fibres has created difficulties for producers of natural fibres. The silk industry was the first to be affected and was almost eliminated by competition from "artificial silk". The cotton and wool industries are still being affected. Whether or not they will eventually suffer to the extent of the silk industry remains to be seen.

The percentage contribution of cotton to the world supply of textile fibres fell from 75 per cent in 1940-41 to 49 per cent in 1976-77. Although cotton still remains the world's main textile fibre it is being increasingly challenged by man-made fibres for this position. The decline in the relative importance of wool as

TABLE 1.1

Percentage World Production Shares of Major Textile Fibres

(based on weight of fibres)

Fibre		1940-41		1960-61		1976-77[a]	
		%		%		%	
Cotton		75.4		67.9		49.3	
Non-Cellulosics	Total Man-Made Fibre	1.0	12.2	4.8	22.3	32.8	45.0
Cellulosics		12.1		17.5		12.2	
Wool (clean)		12.4		9.8		5.7	

Note: (a) Preliminary.

Sources: Based on Table 3.1 in *The Marketing of Australian Wool* [2], and Table II.1 in *The Fibre Review 1975-76* [6].

a textile fibre has been even more marked than that for cotton. Wool's share of the world textile fibre market fell from 12.4 per cent in 1940-41 to 5.7 per cent in 1976-77. Indirectly, this has meant that Australia, which produces 30 per cent of the world's wool (40 per cent of that in non-communist countries), has become less important as a supplier of textile fibres for the world. Since Australia supplies about one-third of the world's wool, it supplied a little over four per cent of the world's requirements of textile fibres in 1940-41 but less than two per cent in 1976-77.

The relative shares of wool, cotton, non-cellulosic and cellulosic fibres in world production of main textile fibres 1960-61 to 1976-77 are detailed in Fig. 1.1.

Despite the fall in the market shares of cotton and wool, the total amount of cotton production and consumption has continued to rise and that of wool has remained stationary in the last 17 years (see Table 1.2). About 24 per cent of the world's increased consumption of textile fibres between 1960-61 and 1976-77 was met from increased supplies of cotton. However man-made fibres were the main source of supply providing for 76 per cent of increased fibre consumption.

The extent of the increased world consumption of textile fibres in the last 15 years or so has only been possible because of the development of man-made fibres. These fibres have ensured cheaper and a greater volume of fibres and have improved world living standards.

An examination of the world per capita consumption of textile fibres (set out in Table 1.3) is instructive. World per capita consumption of textile fibres increased from 5.1 kilograms in

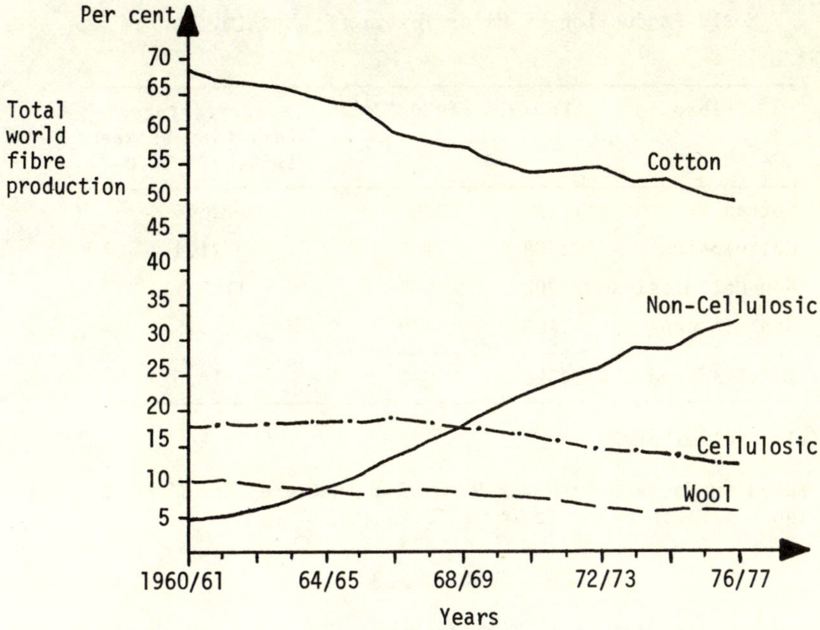

Fig. 1.1

Percentage Share of Main Fibres in Total World Fibre Production 1960-61 to 1976-77 [(a)]

Note: (a) 1975-76 and 1976-77 are preliminary.

Sources: Based on statistics in *Wool Statistics* [18], *Wool Intelligence* [10], *Cotton - World Statistics* [16], and *Textile Organon* [25].

1961 to 6.7 kg. in 1974. Increased use of man-made fibres fully accounted for the per capita rise in consumption. Per capita consumption of cotton rose marginally and that of wool fell to the same extent. Because of increases in the world's population, the world's total consumption of cotton continued to rise and the world's consumption of wool declined only slightly.

In the above discussion, we have referred to two broad types of man-made fibres, cellulosics and non-cellulosics. Cellulosics are produced from plant matter, normally softwood pulp which is processes into small sheets for the industry. Basically, these are dissolved in various chemicals (caustic soda is the main one) to form a liquid which is forced through spinnerets (holes) to form filaments which solidify on leaving the spinneret. Viscose rayon and cellulosic acetate yarns are produced in this way. Thus the cellulosic fibre industry is principally based on a renewable resource.

TABLE 1.2

World Production of Major Textile Fibres (m. kg.)

Fibre	1960-61	1976-77[(a)]	Percentage variation between 1960-61 & 1976-77
Cotton	10113	12802	26.6
Cellulosics	2608	3166	21.4
Non-Cellulosics	702	8509	1112.1
Wool (clean)	1463	1470	0.5
Total Fibres	14886	25947	14.3

Note: (a) Preliminary.

Sources: Based on Table 3.1 in *The Marketing of Australian Wool* [2], and Table II.1 in *Fibre Review 1975-76* [6].

TABLE 1.3

World Consumption of Textile Fibres Per Head (kg./head)

Fibre	1961		1974	
Cotton	3.2		3.3	
Cellulosics	0.9	1.1	0.9	2.8
Non-Cellulosics	0.2		1.9	
Wool	0.5		0.4	
Miscellaneous	0.3		0.2	

Source: *The Fibre Review* [5] (various issues).

The first commercial production[1] of cellulosic fibres was commenced in France by the Chardonnet process in 1885. But this extremely inflammable nitrocellulose fibre was soon replaced by viscose rayon and cuprammonium rayon which were first introduced in Britain and Germany respectively. The products of the

[1] Reviews of the historical growth of the manufacture of man-made fibres and an outline of their methods of production can be found in Airov [1], Cockett [8], Hague [13], Hirsh and Ellis [14], Hufbauer [15], Markham [21], Robson [24], and the ECAFE seminar proceedings published by the Japan Chemical Fibres Association [19].

cellulosic fibres industry rapidly replaced those of the silk industry due principally to the efforts of companies like Courtaulds Ltd. However, in recent years world consumption of cellulosic fibres has tapered off and as Table 1.3 indicates, per capita consumption of these fibres has remained stationary. Man-made cellulosic fibres have experienced considerable competition from non-cellulosic fibres.

The major growth area in the fibre industry, especially since the beginning of the 1960s, has been in the supply of non-cellulosic fibres, produced principally from oil by-products. Nylon, the earliest fibre produced from this source was discovered in the U.S.A. in 1935 by Dr Carothers, a scientest working for the Du Pont chemical company. Du Pont commenced commercial production of nylon not long after. Polyester was discovered in Britain by Whinfield and Dickinson and first manufactured in the U.S.A. by Du Pont in 1949 and in the following year in the United Kingdom by Imperial Chemical Industries (I.C.I.). Du Pont marketed its polyester under the trade name of "Dacron" and I.C.I. used the trade name "Terylene". Du Pont continued to be an outstanding innovator in the synthetic fibre field. In 1950 it commenced production of acrylic fibre ("Orlon") in the U.S.A. The company had discovered this fibre during World War II.

The feed stocks required for the production of oil-based fibres are basically the same as those needed for the production of plastics. The raw ingredients as far as the yarn producer is concerned are polymer chips produced mainly from oil by-products. Yarn producers melt these chips and force the melt through spinneret holes. The molten threads are then cooled and wound at high speed.

The non-cellulosic synthetic fibre industry is based upon the use of *non-renewable resources*. However, the synthetic fibre industry uses a low proportion of total oil production and the raw materials used in the polymer form a low proportion of the total cost of a yarn. For instance, in the case of polyester yarn for apparel, these costs amount to no more than 20 per cent of the expense of the yarn. Substantial rises in the price of oil have little impact on the cost of synthetic yarns. Rising oil prices, as a result of the energy crisis, are likely to have little effect on the cost of producing synthetic yarns [12]. Furthermore, natural gas and coal can be substituted for oil as a basis for making non-cellulosic fibres. While these are not renewable resources, coal is in very abundant supply at present.

The world's consumption and production of fibres have been revolutionised by the advent of man-made fibres. A greater range of fibres is now available with a wider range of properties than ever before, and we have almost reached the stage where the bulk of fibres are man-made. The United Nations estimates that practically all of the world's increase in textile fibre consumption in the next 25 years will be met from man-made fibres. The rapid growth in the world consumption of man-made fibres since the mid-fifties is evident from Fig. 1.2. Whereas consumption of man-made fibres has grown rapidly, consumption of cotton and wool has tapered off and the aggregate level of wool consumption is now stationary or nearly so. Let us look at trends in the production and consumption of man-made fibres, cotton and wool, in more detail.

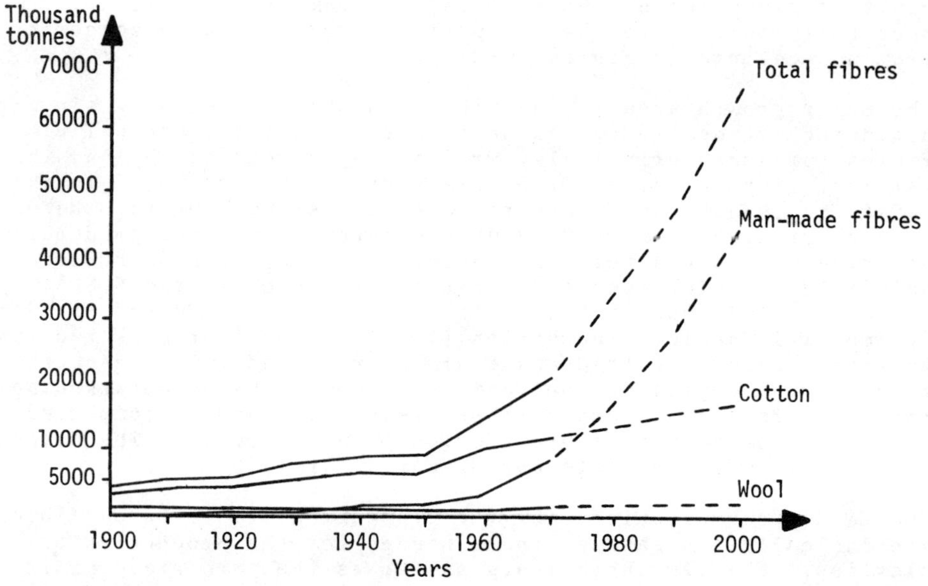

Fig. 1.2

Actual and Projected World Production of Textile Fibres[(a)] 1900-2000[(b)]

(Thousand Tonnes)

Notes: —— Actual Figures.

--- Projected Figures.

(a) Flax is omitted because it is a relatively unimportant fibre.

(b) Data graphed at ten year intervals.

Sources: Actual figures from *Information on Man-Made Fibres* [9] and *Textile Organon* [25]. Projected figures based on estimates found in Table 80 in *An Introduction to Textile Economics* [14].

1.2 MAIN MAN-MADE FIBRES - THEIR PRODUCTION AND CONSUMPTION

If we exclude glass, steel and other metal threads, asbestos and carbon fibres, which principally have industrial uses (non-apparel uses), the main man-made fibres used in apparel and within the home, are those set out in Table 1.4. Note however, that the chain of Triffin-like substitution [26] in the textile industry is a long one and it is difficult to define the limits of the textile fibre industry. This becomes apparent if we consider tyres for instance. Cotton was used in tyres, then various cellulosic fibres, and after this nylon and polyester. Today, steel or fibreglass reinforcement is commonly used. There is considerable scope for substitution between fibres. Furthermore, there are other fibres such as jute, olefins and plastic substitutes, used mainly for packing and carpet-backing, which could be considered in a discussion of the fibre industry. However, since wool is the fabric of most interest to us we shall focus mainly on fibres used in apparel.

TABLE 1.4

Main Man-Made Chemical Fibres

Cellulosic fibres	Non-cellulosic fibres
Viscose rayon	Polyamides (nylon)
Cuprammonium rayon	Polyesters
Cellulosic acetate	Acrylic and modacrylic
Cellulose triacetate	

There are two main groups of man-made fibres - cellulosic fibres and non-cellulosic fibres. The production of cellulosic fibres, as already pointed out, is based upon the use of natural cellulosic material such as wood pulp. The production of non-cellulosic fibres (synthetic or chemical fibres) is based upon the use of plastic-type materials obtained usually from oil or coal. Cellulosic fibres were discovered much earlier than non-cellulosic fibres and, as listed in Table 1.4, there are various common types of each. Let us consider cellulosic fibres and non-cellulosic fibres in turn.

Cellulosic Fibres

In terms of the volume of its use, viscose rayon is the leading cellulosic fibre. Like the other cellulosic fibres, its production is principally based on wood pulp, although other renewable cellulosic material such as cotton linters and bamboo can be used [22]. The viscose rayon industry is now in terms of the theories of Hufbauer [15], Posner [23], and Vernon [28], a mature industry which may be entering the declining stage of its development.

In general, as a result mainly of competition from non-cellulosic fibres, the production of cellulosic fibres has been stationary or declining since 1970 and their percentage share of world fibre production has been declining since 1966 (see Table 1.5 and Fig. 1.3). In addition to this, cellulosic fibres' share of the man-made fibre market has been falling since 1940. Whereas in 1935 cellulosic fibres were the only man-made fibres, by 1976 their share of the man-made fibre market had fallen to 27 per cent. Some large companies, for instance Du Pont de Nemours of the U.S.A. and Teijin of Japan, have discontinued their production of cellulosic fibres presumably because they foresee low profit margins for these fibres and wish to concentrate their capital in areas with higher returns [22].

TABLE 1.5

Production and Percentage Share of Cellulosic Fibres 1960 to 1976[(a)]

Year	Production m. kg.	Percentage of total production
1960	2608	17.6
1961	2690	18.2
1962	2864	18.2
1963	3059	18.2
1964	3286	18.6
1965	3339	18.2
1966	3343	18.8
1967	3315	18.3
1968	3531	17.3
1969	3560	17.2
1970	3433	16.4
1971	3442	15.1
1972	3533	14.3
1973	3666	14.0
1974	3582	13.6
1975	2965	12.5
1976	3166	12.2

Note: (a) 1975 and 1976 figures are preliminary.

Source: Based on statistics in *Textile Organon* [25] (various June issues).

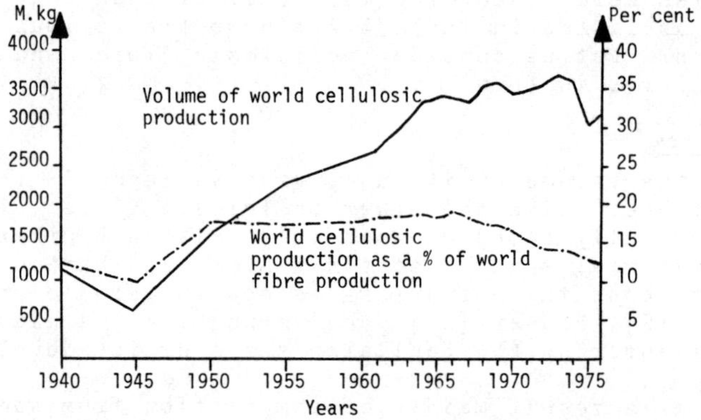

Fig. 1.3

Production and Productive Share of Cellulosic Fibres 1940 to 1976[(a)]

Note: (a) Data from 1940 to 1960 is graphed on a five year basis, and from 1960 on a yearly basis. 1976 figures are preliminary.

Sources: Based on statistics in *The Marketing of Australian Wool* [2], and *Textile Organon* [25] (various June issues).

Production of cellulosic fibre commenced first in Europe, was based on European technology and rapidly spread to the U.S.A. Production today continues to be concentrated in the developed countries of the world with Western Europe, Eastern Europe, U.S.S.R., U.S.A. and Japan the major producing areas (see Tables 1.9 and 1.10). However, less developed countries are increasing their share, a result which Vernon's theory [28] predicts for a mature industry.

The share of the major producers of cellulosic fibre in world production in 1975 is shown in Fig. 1.4. Japan has increased its share of world cellulosic fibre production considerably in the last two decades and in 1976 its share was just below that of the U.S.A., the share of which is declining. U.S.S.R. is the world's leading single producing country.

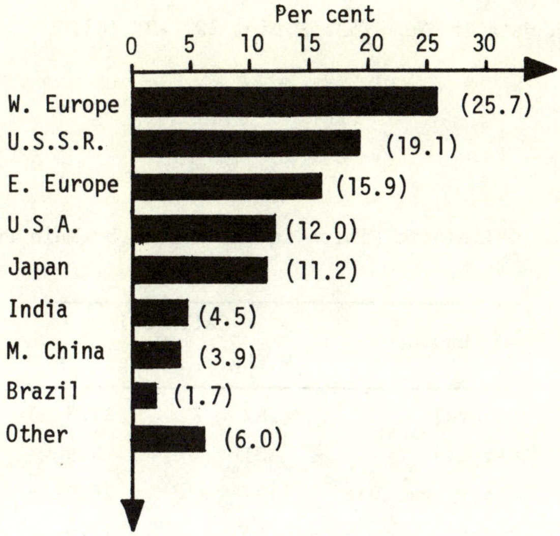

Fig. 1.4

Share of Main Cellulosic Fibre Producing Areas in World Production 1976

Source: Based on figures in *Textile Organon* [25].

Cellulosic fibres have a wide range of end-uses. they are used for clothing, household items and in industry. Tables 1.6 and 1.7 indicate that considerable quantities of cellulosic fibres are consumed in each of these end-uses. However, in each of these areas man-made cellulosic fibres have experienced considerable competition from non-cellulosic fibres. For instance, as Table 1.6 indicates cellulosic fibres have found considerably reduced consumption in all of their end-uses in the U.S.A. since 1970.

TABLE 1.6

Consumption of Cellulosic Fibres in the U.S.A. by Main End-Uses 1970 and 1975

End-use	1970 M. kg.	1970 Per cent	1975 M. kg.	1975 Per cent	Percentage change in consumption between 1970 & 1975
Apparel	317	50.2	156	44.5	-50.8
Home furnishings	168	28.6	84	23.9	-50.0
Industrial, etc.	147	23.2	111	31.6	-24.5
Total	632	100.0	351	100.0	-44.5

Source: Based on data in *The Fibre Review 1975-76* [6].

TABLE 1.7

Consumption of Cellulosic Fibres in the E.E.C. by Main End-Uses 1975

End-use	Cellulosic M. kg.	Cellulosic Per cent
Apparel	192	37.3
Carpets	31	5.9
Other home use	137	26.7
Tyres	66	12.9
Other technical	89	17.2
Total	515	100.0

Source: Based on Table II in "Man-Made Fibres and Natural Fibres: Competition in End-Uses in Western Europe" [3].

Non-Cellulosic Fibres

As Table 1.8 and Fig. 1.5 indicate world production of non-cellulosic fibres expanded rapidly during the 1960s and early 1970s. By 1976 non-cellulosic fibres accounted for 73 per cent of world production of man-made fibres and 33 per cent of all fibres produced. In the decade to 1976 production of synthetic fibres increased eightfold.

TABLE 1.8

Production and Percentage Share of Non-Cellulosic Fibres 1960 to 1976[(a)]

Year	Production m. kg.	Percentage of total production
1960	639	4.3
1961	752	5.1
1962	981	6.2
1963	1232	7.4
1964	1632	9.2
1965	1975	10.7
1966	2370	13.3
1967	2730	15.2
1968	3578	17.6
1969	4178	20.3
1970	4693	22.5
1971	5595	24.5
1972	6342	25.7
1973	7659	28.8
1974	7455	28.4
1975	7363	31.1
1976	8509	32.8

Note: (a) 1975 and 1976 figures are preliminary.

Source: Based on statistics in *Textile Organon* [25] (various June issues).

The main non-cellulosic textile fibres are the polyamides (nylons), polyesters and acrylics, including modified acrylics.[2] Commercial production of polyamides (nylons) commenced during World War II and commercial production of polyester and acrylic fibres started at the beginning of the 1950s. All are produced mainly from products of oil or natural gas but coal can be substituted as a basis for their production.

Polyamides (nylons) are produced by first manufacturing a nylon salt which is melted, extruded in a ribbon and broken into chips. These "plastic" chips are then used for spinning nylon filaments. The chips are heated electrically, the melt filtered and spun from spinnerets. As the melt leaves the spinnerets, it is cooled by the air and forms filaments which may be used to produce yarn.

Nylon was discovered in the U.S.A. in 1935 and a little later in Germany. Whereas Nylon 66 was the type of nylon adopted for production in the U.S.A., Nylon 6 was the polyamide adopted at first in Germany. The properties of these polyamides are slightly different. The largest proportion of world production of polyamides today is of Nylon 66 but world production of Nylon 6 is not insignificant.

---

[2] A nomenclature and complete commercial list of man-made fibres can be found in Moncrieff's book *Man-Made Fibres* [22, pp. 10-11; 1043-1064].

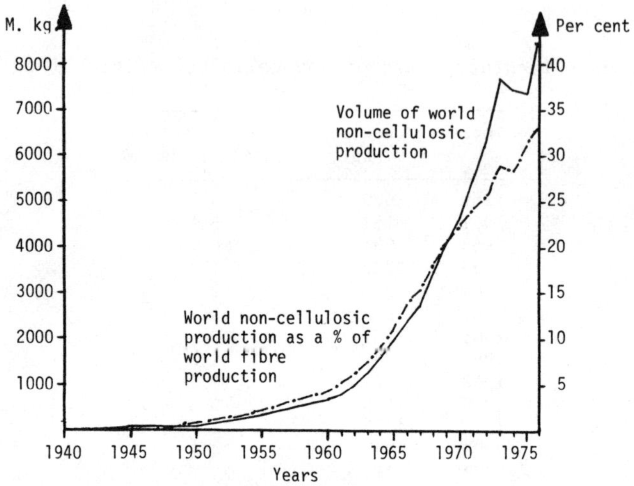

Fig. 1.5

World Production of Non-Cellulosic (Synthetic) Fibres 1940 to 1976[(a)]

Note: (a) Data from 1940 to 1960 graphed at five year intervals and from 1960 in yearly intervals. 1976 figures are preliminary.

Sources: Based on statistics in *The Marketing of Australian Wool* [2], *The Fibre Review 1975-76* [6] and *Textile Organon* [25] (various issues).

World production of nylon expanded rapidly during the 1950s and especially the 1960s, see Fig. 1.6, but in recent years expansion has slowed down and the nylon's share of non-cellulosic fibre sales has fallen as is indicated in Fig. 1.7.

Polyester was discovered in England by Whinfield and Dickinson of the Calico Printers' Association following up Carothers' work. Soon after its commercial production commenced in 1950, polyester began to capture a major share of the non-cellulosic fibre market (see Fig. 1.7). Figs 1.6 and 1.7 indicate that consumption of polyester has, since 1950, grown at the fastest rate of all man-made fibres and it has consistently increased its share of the synthetic fibre market. It has maintained its high growth rate in the face of falling growth rates for nylon and acrylic.

At the present time, polyester is manufactured mostly from oil derivatives which are processed to form polymer chips. Polyester fibre is formed by melting these chips and forcing the molten polymer through a spinneret. The air solidifies the individual filaments almost instantly as they leave the spinneret and the filaments are drawn together to form yarn and wound on to cylinders.

Production of acrylic fibres commenced at much the same time as that of polyesters but the initial spread in their use was slower.

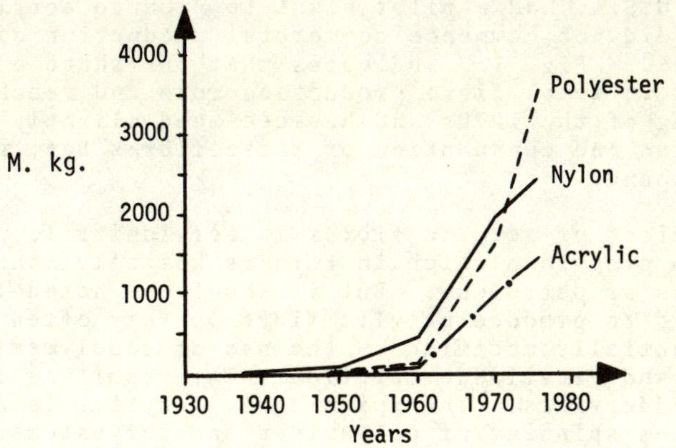

Fig. 1.6

World Production of Polyamides, Polyesters and Acrylics 1930 to 1975[(a)]

Note: (a) Data graphed at ten year intervals up to 1970 and then at five year intervals to 1975.

Sources: Based on statistics in *An Introduction to Textile Economics* [14] and *Textile Organon* [25] (various issues).

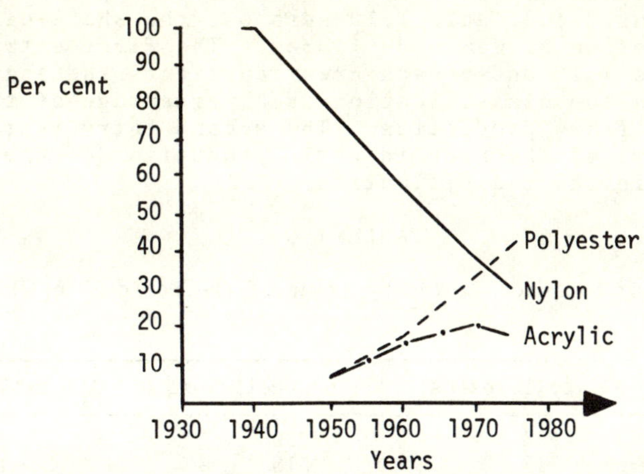

Fig. 1.7

Shares of Polyamides, Polyesters and Acrylics in Non-Cellulosic Fibre Production 1930 to 1975[(a)]

Note: (a) Data graphed at ten year intervals up to 1970 and then at five year intervals to 1975.

Sources: Based on statistics in *An Introduction to Textile Economics* [14], and *Textile Organon* [25] (various issues).

Du Pont of the U.S.A. had a pilot plant to produce acrylic fibres in 1945 but it did not commence commercial production of these fibres until 1950. Fig. 1.7 indicates that the share of acrylics and modacrylics in total fibre production rose and reached a peak by the beginning of the 1970s and has declined slightly since then. Production and consumption of these fibres has, however, continued to expand.

The main ingredient of acrylic fibres is acrylonitrile produced, as a rule, from propylene which in turn is normally manufactured from natural gas or petroleum. But it should be noted that coal can also be used to produce acrylic fibres. Very often acrylic fibre is substantially modified by the use of copolymers such as vinyl chloride and vinylidene chloride. The resulting fibres are described as modacrylics. The spinning of acrylics is more complicated than the spinning of polyamides and polyesters. The compound to be used for spinning must be dissolved in solvents.

Petrochemical companies produce the main ingredients needed for the production of synthetic fibres and it is, therefore, not surprising that many of these companies have since World War II begun to produce man-made fibres. Chemical companies such as Union Carbide and Imperial Chemical Industries for example, commenced production of synthetic fibres after World War II.

Table 1.9 indicates the level of production of man-made fibres (cellulosic and non-cellulosic) of each of the main producing areas of the world and Table 1.10 sets out the share of each area in world production of man-made fibres. The first entry in each column of Table 1.10 under each area represents the share of the fibre listed in the classification as a percentage of that area's total man-made fibre production. The second entry represents the percentage share of the area in world production of the corresponding fibre in the classification.

TABLE 1.9

Production of Man-Made Fibres by Areas of the World 1976 (M.kg.)

Producing areas	All cellulosics	All non-cellulosics	All man-made fibres
Western Europe	826	2299	3125
Eastern Europe	512	515	1027
U.S.S.R.	612	405	1017
U.S.A.	385	2745	3130
Japan	360	1204	1564
Korea	14	309	323
Other Americas	142	544	686
All Other	362	577	939
Total	3212	8598	11810

Source: Based on data in *Textile Organon* [25].

TABLE 1.10

Production Share[(a)] of Man-Made Fibres of the Main Producing Areas of the World 1976

Producing areas	All cellulosics	All non-cellulosics	All man-made fibres
Western Europe	(25.7, 26.4)	(26.7, 73.6)	(26.3, 100.0)
Eastern Europe	(15.9, 44.4)	(6.0, 50.1)	(8.7, 100.0)
U.S.S.R.	(19.0, 60.2)	(4.7, 39.8)	(8.6, 100.0)
U.S.A.	(12.0, 12.3)	(31.9, 87.7)	(26.5, 100.0)
Japan	(4.2, 23.0)	(14.0, 77.0)	(13.3, 100.0)
Korea	(0.5, 4.3)	(3.7, 95.7)	(2.9, 100.0)
Other Americas	(4.4, 20.2)	(6.3, 79.3)	(5.8, 100.0)
All Other	(11.3, 38.6)	(6.7, 61.4)	(7.9, 100.0)
Total	(100.0, 27.2)	(100.0, 72.8)	(100.0, 100.0)

Note: (a) The first entry in brackets in each row for each area represents the percentage share of the area in world production of the corresponding fibre in the classification. The second entry represents the share of the fibre listed in the classification as a percentage of that area's total man-made fibre production.

Source: Based on Table 1.9.

It is interesting to note differences in the degree to which different areas of the world specialise in non-cellulosic fibre production. This is readily seen from Fig. 1.8. Korea and the U.S.A. have specialised to the greatest extent in non-cellulosic fibre production whereas the U.S.S.R. and Eastern Europe have specialised to the greatest extent in cellulosic fibre production.

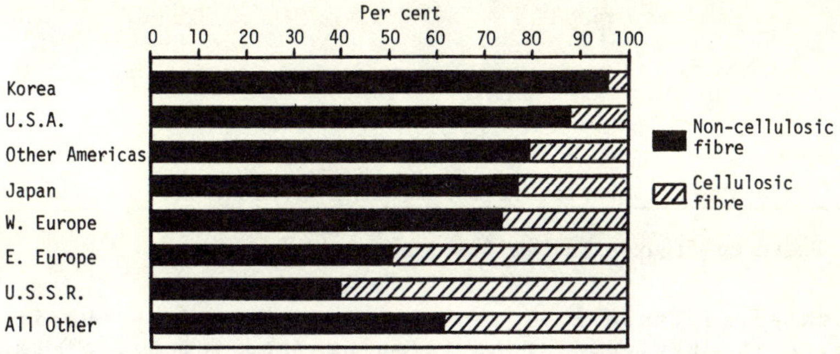

Fig. 1.8

Degree of Specialisation in Non-Cellulosic Fibre Production (in comparison to cellulosic production) in the Main Man-Made Fibre Producing Areas of the World 1976

Source: Based on Table 1.10.

The world's production of synthetic fibres is chiefly located in major developed countries most of which rely on imported materials (oil) to produce these fibres. As discussed by Hufbauer [11] technological leads, economies of scale and trade barriers may help to explain this concentration. In terms of technological progress in the man-made fibre industry, Western Europe and the U.S.A. have led the world. Producers in both areas have had substantial home markets to provide initial economies of scale.

More than 90 per cent of the world's production of synthetic fibres is still undertaken in developed countries even though the share of less developed countries has risen. The proportion of the world's production accounted for by selected areas in indicated in Fig. 1.9.

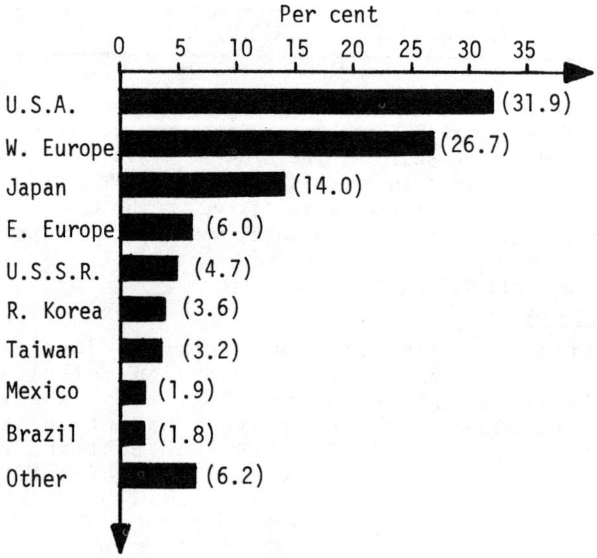

Fig. 1.9

Share of Main Non-Cellulosic Fibre Producing Areas in World Production 1976

Source: Based on figures in *Textile Organon* [25].

As indicated in Table 1.11 the concentration of synthetic fibre production in developed areas is maintained for all of the major non-cellulosic fibres - polyamides, polyesters and acrylics. Table 1.11 also indicates that the degree of specialisation in different types of non-cellulosic fibres tends to vary between areas.

Differences in the degree of specialisation in producing different types of non-cellulosic fibres can be readily seen from

TABLE 1.11

Production of the Major Non-Cellulosic Fibres by Areas of the World and Percentage Shares[(a)] of the Areas in the Production 1976 (m. kg.)

Producing areas	Polyester		Polyamide		Acrylic		*Total*	
	Vol.	%	Vol.	%	Vol.	%	Vol.	%
Western Europe	800	(20.5, 35.2)	751	(26.5, 33.1)	721	(41.6, 31.7)	2272	(26.8, 100.0)
Eastern Europe[(b)]	337	(8.6, 37.3)	391	(13.8, 43.3)	175	(10.1, 19.4)	903	(10.7, 100.0)
U.S.A.	1515	(38.8, 55.3)	941	(33.3, 34.4)	282	(16.3, 10.3)	2738	(32.3, 100.0)
Japan	535	(13.7, 46.7)	301	(10.6, 26.3)	310	(17.9, 27.0)	1146	(13.5, 100.0)
Other Americas	266	(6.8, 49.0)	195	(6.9, 35.9)	82	(4.7, 15.1)	543	(6.4, 100.0)
All Other	454	(11.6, 52.4)	251	(8.9, 29.0)	162	(9.4, 18.6)	867	(10.3, 100.0)
Total	3907	(100.0, 46.1)	2830	(100.0, 33.4)	1732	(100.0, 20.5)	8469	(100.0, 100.0)

Note: (a) The first entry in brackets in each row under each fibre represents the percentage share of the area in world production of the corresponding non-cellulosic fibre in the classification. The second entry represents the share of the non-cellulosic fibre listed in the classification as a percentage of that area's total non-cellulosic production.

(b) Includes U.S.S.R.

Source: Based on data in *Textile Organon* [25].

Fig. 1.10. Western Europe and Japan show a notably higher concentration of production in acrylic production than the U.S.A. and the volume of both Western Europe and Japanese production of acrylics even exceeds that of the U.S.A. Eastern and Western Europe concentrate to a greater extent than either the U.S.A. or Japan on polyamide production even though the U.S.A. is the largest producer of polyamide in the world. The U.S.A. shows the greatest specialisation of any area in the world in polyester production and is by far the largest producer of this fibre in the world.

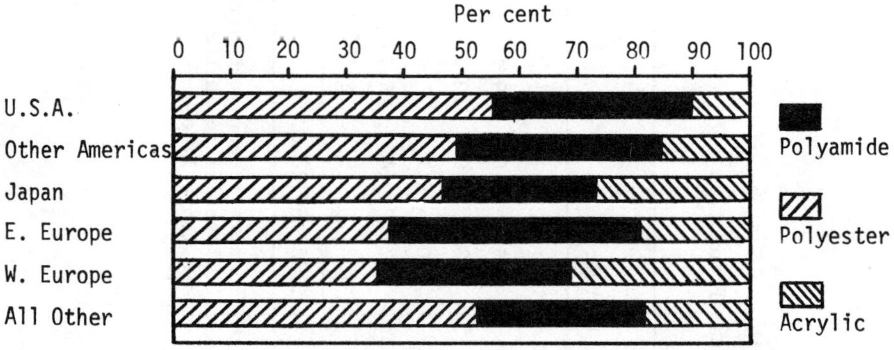

Fig. 1.10

Degree of Specialisation in Production of Main Non-Cellulosic Fibres by Major Producing Areas of the World 1976

Source: Based on Table 1.11.

Returning to the question of the geographical distribution of production of the main synthetic fibres of the world it is clear from Table 1.11 and Figs 1.11, 1.12 and 1.13 that it is heavily concentrated in the developed areas of the world. Although there are some differences for some fibres, the U.S.A. is the dominant world producer of non-cellulosic fibres, followed by Western Europe, Japan and Eastern Europe (including U.S.S.R.).

In the case of polyester, the principal synthetic fibre, U.S.A. *dominates* world production. It accounts for almost 40 per cent of world production and its output is almost twice as great as that of Western Europe, the next most considerable producer. Japan's share of world production of polyester is considerable (see Fig. 1.11) especially so in view of its relatively smaller population compared to other areas. The rather low relative proportion for Eastern Europe possibly reflects the lag in technology transfer to this area. The 'Big Four' (U.S.A., Western Europe, Japan and Eastern Europe) account for over 80 per cent (81.6%) of world polyester production.

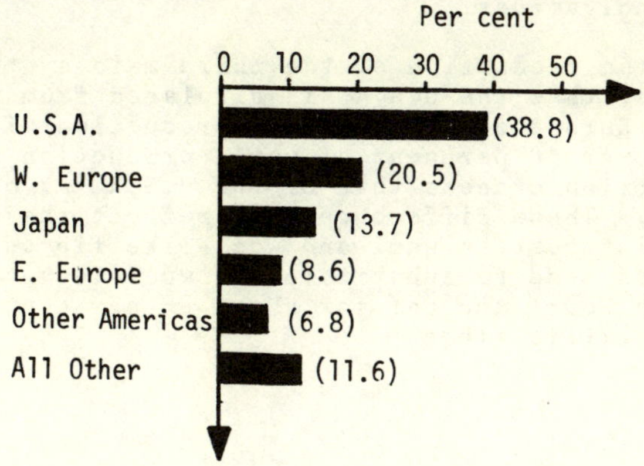

Fig. 1.11

Share of Main Areas in World Polyester Production 1976

Source: Based on Table 1.11.

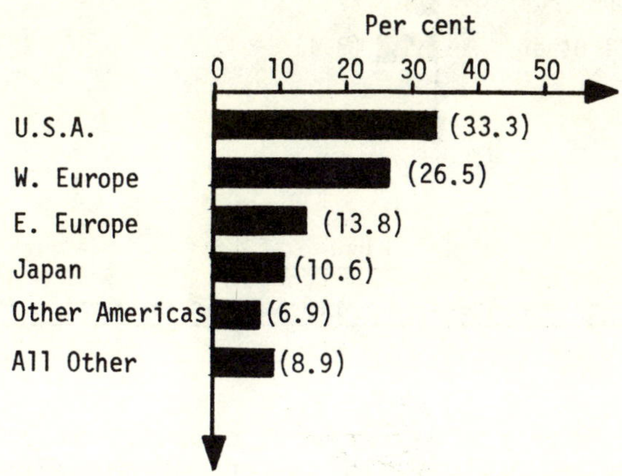

Fig. 1.12

Share of Main Areas in World Polyamide Production 1976

Source: Based on Table 1.11.

The U.S.A. produces about one-third of the world's output of polyamide and is the world's largest producer of this fibre. However, Eastern Europe production is nearly as great as that of the U.S.A. The relative shares of the major areas in world polyamide production is graphed in Fig. 1.12. The 'Big Four' (U.S.A., Western

Europe, Eastern Europe, Japan) account for 86.2 per cent of world production of polyamides.

It is only in the production of the third main synthetic type of fibre, acrylics, that the U.S.A. is displaced from the first rank. Western Europe dominates world production of acrylics. It accounted for over 40 per cent of world production in 1976. Even Japanese production exceeds that of the U.S.A. as can be seen from Fig. 1.13. These differences may reflect the apparent relative greater interest in wool and wool-like fibres in Europe and Japan. Acrylics tend to substitute for wool in a number of end-uses. The 'Big Four' account for about 86 per cent of world production of acrylic fibres.

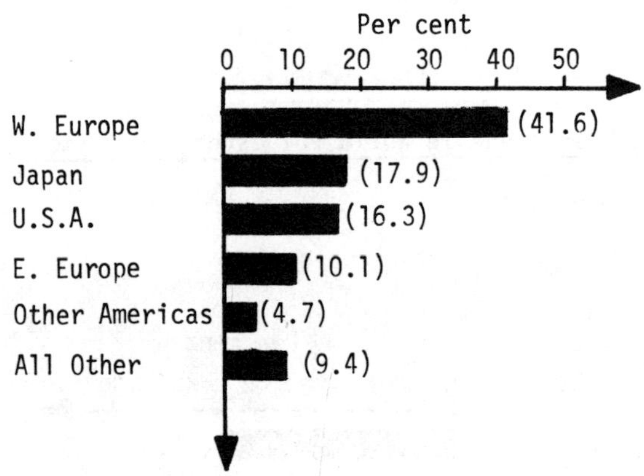

Fig. 1.13

Share of Main Areas in World Acrylic Production 1976

Source: Based on Table 1.11.

Non-cellulosics have a wide range of uses in the clothing, household and industrial fields. The extent of their use in these various fields in the U.S.A. and the E.E.C. is indicated in Tables 1.12 and 1.13. The principle use of non-cellulosic fibres is in apparel, and more than half of their total use in the U.S.A. and in the E.E.C. is in apparel and home furnishings. The consumption of non-cellulosic fibres has expanded in all of their end-uses.

It is also interesting to consider the consumption of non-cellulosic fibres by regions. Excluding communist countries over 80 per cent of non-cellulosic staple and more than three-quarters of non-cellulosic filament yarn is consumed in developed countries. The United States and Western Europe alone account for 70 per cent of staple consumption and 61 per cent of filament consumption. The regional distribution of consumption of non-cellulosic fibres is summarised in Table 1.14.

TABLE 1.12

Consumption of Non-Cellulosic Fibres in the U.S.A. by Main End-Uses 1970 and 1975

End-use	1970		1975		Percentage change in consumption between 1970 & 1975
	M. kg.	Per cent	M. kg.	Per cent	
Apparel	751	41.0	1250	42.4	+66.5
Home furnishings	595	32.5	970	33.0	+63.0
Industrial, etc.	484	26.5	723	24.6	+49.4
Total	1830	100.0	2943	100.0	+60.8

Source: Based on data in *The Fibre Review 1975-76* [6].

TABLE 1.13

Consumption of Non-Cellulosic Fibres in the E.E.C. by Main End-Uses 1975

End-Use	Non-cellulosic	
	M. kg.	Per cent
Apparel	773	55.4
Carpets	274	19.6
Other home use	198	14.2
Tyres	27	1.9
Other technical	124	8.9
Total	1396	100.0

Source: Based on Table II in "Man-Made Fibres and Natural Fibres: Competition in End-Uses in Western Europe" [3].

TABLE 1.14

Percentage of Non-Communist World Consumption of Non-Cellulosic Fibres in Geographical Areas of the World

Producing area	Synthetic staple		Synthetic filament	
United States	39.0		38.0	
Western Europe	31.0		23.0	
Japan	8.0		10.0	
Canada	2.0		3.0	
Australia, New Zealand	0.7		0.8	
Africa, Asia, Oceania (developed)	0.6		1.3	
Total developed areas		81.0		77.6
South America	5.6		8.0	
Africa, Asia, Oceania (less developed)	13.4		14.4	
Total less developed areas		19.0		22.4
TOTAL		100.0		100.0

Source: Based on statistics in "The Present World Market for Polyester Fibres" [20].

1.3 PATTERNS IN COTTON PRODUCTION AND CONSUMPTION

Cotton is still the world's main textile fibre and accounts for 49 per cent of the volume of textile fibres produced in the world (see Table 1.15 and Fig. 1.14). However its share of world consumption of textile fibres has been falling partly because of competition from man-made fibres and partly because man-made fibres have expanded the range of uses of fibres.

World per capita consumption of cotton has declined since 1960 after reaching its maximum then, whereas per capita consumption of man-made fibres has been rising. Although cotton accounted for almost three-quarters of total textile fibre consumption at the beginning of the 1950s, by the mid-1970s its share had fallen to 50 per cent.

The U.S.A. is no longer the world's leading producer of cotton, although it still is an important producer. Its relative importance as a cotton producer has declined and its production is exceeded by that of the U.S.S.R. In 1947-49 the United States produced 48 per cent of the world's cotton but by 1975-76 this share had fallen to 19 per cent of the world's output of cotton. Production of cotton by underdeveloped countries has expanded at

TABLE 1.15

Production and Percentage Share of Cotton 1960-61 to 1976-77[(a)]

Year	Production m. kg.	Percentage of total production
1960-61	10113	68.2
1961-62	9819	66.6
1962-63	10457	66.2
1963-64	10955	65.4
1964-65	11308	63.8
1965-66	11604	63.0
1966-67	10556	54.2
1967-68	10411	57.8
1968-69	11623	57.1
1969-70	11268	54.7
1970-71	11171	53.5
1971-72	12235	53.6
1972-73	13362	54.1
1973-74	13616	51.8
1974-75	13777	52.4
1975-76	11881	50.1
1976-77	12802	49.3

Note: (a) 1975-76 and 1976-77 figures are preliminary.

Source: Based on data in *Cotton - World Statistics* [16] (various issues).

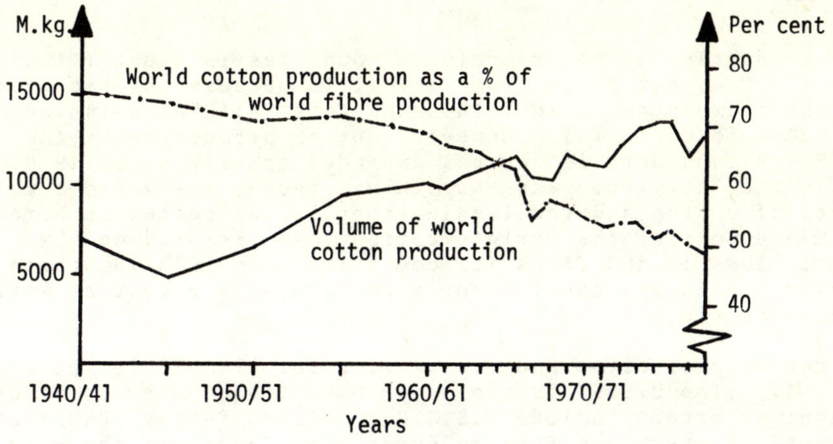

Fig. 1.14

World Production of Cotton 1940-41 to 1976-77[(a)]

Note: (a) Data from 1940 to 1960 graphed at five year intervals and from 1960 in yearly intervals. 1976-77 figures are preliminary.

Sources: Based on statistics in *The Marketing of Australian Wool* [2], and *The Fibre Review 1975-76* [6].

a faster rate than in developed countries. In 1970, as Table 1.16 indicates, these areas produced a considerable portion of the world's supply of cotton fibre. In 1975-76 less developed countries produced more than half of the world's output of cotton.

TABLE 1.16

Shares of Economic Regions in World Raw Cotton Production 1970

Economic class of region	Volume of production in '000 tons	Production as a percentage of world production
Economic Class I[(a)] (developed)	2943	24.1
Economic Class II (underdeveloped)	5441	44.5
Economic Class III[(b)] (centrally planned)	3845	31.4

Notes: (a) Includes U.S.A. production of 2684 thousand tons.

(b) Includes Chinese production of 1690 thousand tons.

Source: Based on data in Vol. II of *Agricultural Commodity Projection 1970-1980* [11, p. 248].

Cotton production by underdeveloped countries is greater than it may at first appear to be from Table 1.16 because Chinese production, which amounted to 1690 thousand tons in 1970 is included under Economic Class III. However, cotton production by the U.S.S.R. is considerable and has expanded greatly since 1970. In fact, in the last four years Russia has become the world's major producer of cotton and the leading supplier of cotton in Europe [7]. Hence most of the world's cotton today is produced by Economic Class II and Class III countries. In 1975-76, these countries together accounted for more than 80 per cent of world cotton production.

World cotton production and consumption for 1975-76 is set out in Table 1.17. The U.S.A. is the major exporting country. Other important exporters include U.S.S.R., Mexico, Turkey, Pakistan and Egypt. Western and Eastern Europe and Japan are the main net importing areas. While there is considerable international trade in cotton nevertheless less than one-third of the world's cotton production enters international trade compared to over 45 per cent for world production of wool.

The comparative shares of the main cotton producing countries of the world in the world's total production of cotton is shown in Fig. 1.15. In terms of their relative contribution the major producers are the U.S.S.R., closely followed by China, the U.S.A., India, Pakistan, Turkey, Brazil and the Arab Republic.

TABLE 1.17

World Cotton Production and Consumption 1975-1976

Producing area and country	Production M. kg.	Production Per cent	Consumption[(a)] M. kg.	Consumption[(a)] Per cent
Asia, Oceania	4980	41.9	6757	48.7
U.S.S.R., Eastern Europe	2663	22.4	2648	19.1
North America	278	19.2	1889	13.6
Africa	996	8.4	562	4.0
South America	787	6.6	701	5.1
Western Europe	177	1.5	1320	9.5
Total	11881	100.0	13877	100.0
U.S.A.	1808		1583	
Mexico	197		184	
Brazil	396		412	
China	2385		2775	
India	1132		1312	
Pakistan	514		477	
Turkey	481		293	
Japan	-		694	
Egypt, Arab Republic	382		228	

Note: (a) Preliminary.

Source: Based on Tables II.7 and III.14 in *The Fibre Review 1975-76* [6].

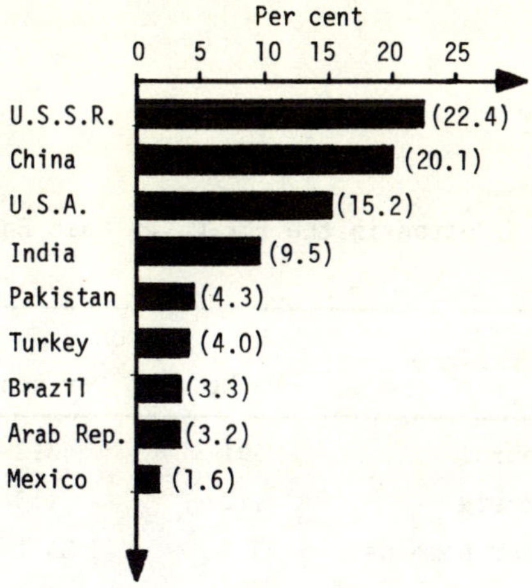

Fig. 1.15

Percentage of World Cotton Production in 1975-76 Accounted for by the Main Cotton Producing Countries

Source: Based on Table 1.17.

Unlike wool which finds comparatively little industrial application, cotton finds widespread use in clothing, household and industrial fields. However, in many of its industrial applications, cotton has been displaced by man-made fibres. In the U.S.A. and E.E.C. three-quarters of cotton consumption is now for apparel and home use (see Tables 1.18 and 1.19). Table 1.18 indicates that cotton consumption in the U.S.A. declined in all of its main end-uses between 1970 and 1975. This decline is to a considerable extent due to increased competition from man-made fibres.

TABLE 1.18

Consumption of Cotton in the U.S.A. by Main End-Uses 1970 and 1975

End-use	1970		1975		Percentage change in consumption
	M. kg.	Per cent	M. kg.	Per cent	between 1970 & 1975
Apparel	819	47.8	732	52.7	-10.6
Home furnishings	570	33.3	410	29.5	-28.1
Industrial, etc.	323	18.9	246	17.8	-23.8
Total	1712	100.0	1388	100.0	-18.9

Source: Based on data in *The Fibre Review 1975-76* [6].

TABLE 1.19

Consumption of Cotton in the E.E.C. by Main End-Uses 1975

End-use	Cotton	
	M. kg.	Per cent
Apparel	391	43.2
Carpets	12	1.3
Other home use	318	35.1
Tyres	6	0.6
Other technical	178	19.8
Total	905	100.0

Source: Based on Table II in "Man-Made Fibres and Natural Fibres: Competition in End-Uses in Western Europe" [3].

1.4 PATTERNS IN WOOL PRODUCTION AND CONSUMPTION

Wool, although it accounts for a small proportion of fibre production, is an important fibre, the qualities of which have not been exactly replicated by a man-made fibre. It is by far the most important animal fibre used in the textile industry.

Wool has experienced considerable competition from man-made fibres. Its share of total textile production has been declining since at least 1945 and now constitutes less than six per cent of total fibre production (as indicated by Table 1.20 and Fig. 1.16). Total volume of wool production now appears to be slowly declining.

TABLE 1.20

Production and Percentage Share of Wool 1960-61 to 1976-77[(a)]

Year	Production m. kg.	Percentage of total production
1960-61	1463	9.9
1961-62	1488	10.1
1962-63	1477	9.4
1963-64	1506	9.0
1964-65	1480	8.4
1965-66	1491	8.1
1966-67	1553	8.7
1967-68	1574	8.7
1968-69	1620	8.0
1969-70	1605	7.8
1970-71	1590	7.6
1971-72	1583	6.8
1972-73	1453	5.9
1973-74	1418	5.4
1974-75	1477	5.6
1975-76	1497	6.3
1976-77	1470	5.7

Note: (a) 1975-76 and 1976-77 figures are preliminary.

Sources: Wool Statistics [18], and *Wool Intelligence* [10] (various issues of each).

The production of wool and the production of wool textiles are carried out on the whole in areas which are extremely distant from one another if production by communist countries is excluded. More than half of the world's wool is produced in the Southern Hemisphere and shipped to developed countries in the Northern Hemisphere for processing. Australia is the world's chief producer of wool, followed by the U.S.S.R. The main producers of raw wool in 1976-77 are listed in Table 1.21. Table 1.21 lists

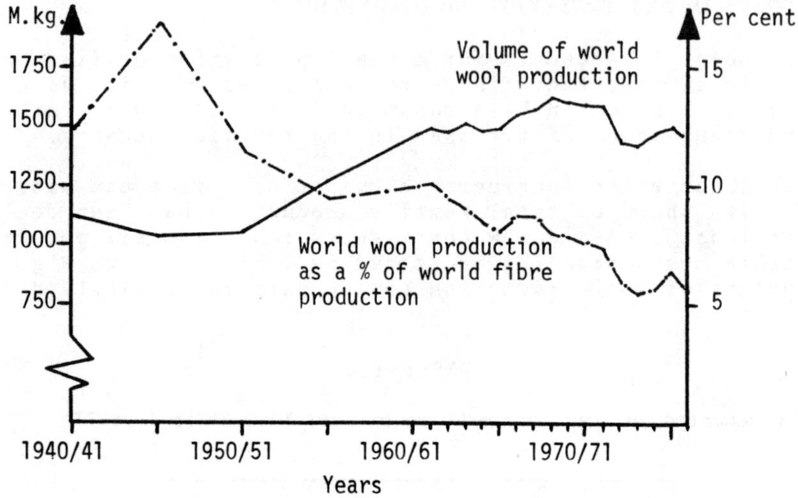

Fig. 1.16

World Production of Wool 1940-41 to 1976-77

Note: (a) Data from 1940 to 1960 graphed at five year intervals and from 1960 in yearly intervals. 1976 figures are preliminary.

Sources: Based on statistics in *The Marketing of Australian Wool* [2] and *The Fibre Review 1975-76* [6].

TABLE 1.21

World Wool Production 1976-1977[(a)] by Countries

Country	Million kg. greasy	Percentage of world production
Australia	725	28.4
U.S.S.R.	450	17.7
New Zealand	312	12.2
Argentina	190	7.5
South Africa	108	4.2
Eastern Europe	95	3.7
China	81	3.2
Uruguay	62	2.4
U.S.A.	53	2.1
United Kingdom	47	1.8
Other countries	425	16.8
World Total	2548	100.0

Note: (a) Preliminary.

Source: Based on data in Table II.2 in *The Fibre Review 1975-76* [6].

the level of their production in 1976-77 and their share of world production. Fig. 1.17 enables the relative contribution of the main producers to world production to be easily compared.

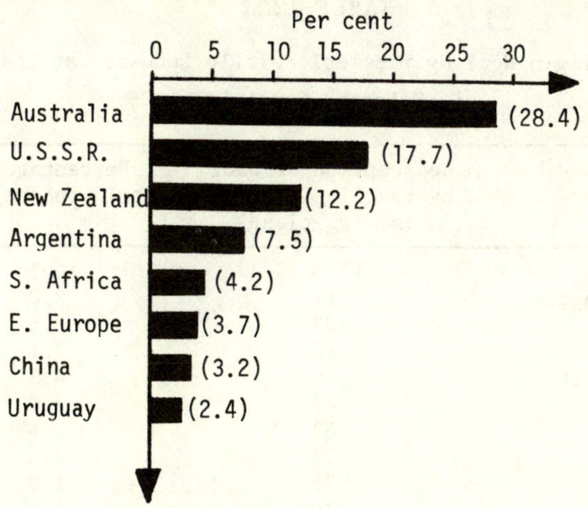

Fig. 1.17

Percentage of World Wool Production Accounted for by Main Producing Countries in 1976-77

Source: Based on Table 1.21.

Australia, New Zealand, Argentina, South Africa and Uruguay account for more than half of the world's output of wool (or almost three-quarters of that in non-centrally planned economies). Each of these countries exports most of its greasy wool to distant areas, mostly developed countries in the Northern Hemisphere, although Argentina and Australia each has a small wool textile industry. At present, Australia exports 90 per cent of its greasy wool production.

Apart from the centrally planned economies which also import substantial quantities of wool, nations producing the largest volume of wool textiles rely heavily on imports of wool. Outside the centrally planned economies, the main wool textile producers are Japan, United Kingdom, France and Italy.

Table 1.22 lists the amount of virgin wool used in 1975 by the wool textile industries in the different countries of the world. An interesting feature, contrasting with its importance in the production of man-made and cotton textiles, is the relative unimportance of the U.S. in the production of wool textiles. From

Table 1.22 it is apparent that the world's production of wool textiles is concentrated in Europe and Japan.

TABLE 1.22

Consumption of Virgin Wool by the Wool Textile Industry at the Carding Stage in Selected Countries 1975

Country	Consumption of wool by textile industry (m. kg. clean)	Percentage of world consumption
Japan	133	10.7
United Kingdom	111	9.0
France	107	8.6
Italy	88	7.1
Germany, F.R.	56	4.5
U.S.A.	48	3.9
Argentina	32	2.6
Turkey	28	2.3
Spain	23	1.9
Australia	20	1.6
U.S.S.R., China, Eastern Europe	476	38.4
Others	116	9.4
World Total	1238	100.0

Source: Based on Table III.2 in *The Fibre Review 1975-76* [6].

On the whole wool is a specialty fibre most of which is used for producing apparel and other clothing. Smaller quantities are used in home furnishings such as carpets and very small amounts are employed for industrial purposes. While the relative quantity of wool used in most textile end-uses has tended to decline, the decline has not been so rapid in apparel and clothing end-uses as in household end-uses, for instance, carpets. In commenting on the position observed up to 1974-75, the Australian Bureau of Agricultural Economics said:

> Over the past decade wool's competitive position has been progressively weakened through the inroads made upon its end uses by man-made fibres. Even in worsted suiting, one of the few end use areas in which wool is the dominant fibre, competitive pressure from non-cellulosic fibres is increasing [5, p. 12].

The consumption of wool in its main end-uses in the U.S.A. and E.E.C. in 1975 is shown in Tables 1.23 and 1.24. These Tables indicate that approximately 70 per cent of wool is used in apparel. Like the consumption of cotton, wool consumption declined in all of its main end-uses in the U.S.A. between 1970 and 1975.

TABLE 1.23

Consumption of Wool in the U.S.A. by Main End-Uses 1970 and 1975

End-use	1970		1975		Percentage change in consumption between 1970 & 1975
	M. kg.	Per cent	M. kg.	Per cent	
Apparel	108	70.6	43	71.7	-60.2
Home furnishings	41	26.8	14	23.3	-65.9
Industrial, etc.	4	2.6	3	5.0	-25.0
Total	153	100.0	60	100.0	-60.8

Source: Based on data in *The Fibre Review 1975-76* [6].

TABLE 1.24

Consumption of Wool in the E.E.C. by Main End-Uses 1975

End-use	Wool	
	M. kg.	Per cent
Apparel	268	69.7
Carpets	77	19.9
Other home use	28	7.3
Tyres	-	-
Other technical	12	3.1
Total	385	100.0

Source: Based on Table II in "Man-Made Fibres and Natural Fibres: Competition in End-Uses in Western Europe" [3].

The consumption of wool in apparel is relatively much more important as an end-use for wool than for any other fibre (see Table 1.25). While the relative utilisation of wool consumption in home furnishings and other home uses is less than for the other fibres, this end-use still remains a substantial outlet for wool accounting for one-quarter of its use compared to one-quarter to one-third of the consumption of other fibres. Wool, compared to all other fibres, obtains little or no use in industrial and other applications, such as for tyre manufacture.

TABLE 1.25

Comparative Allocation of Main Textile Fibres (Percent) to their Principal Uses in U.S.A. and E.E.C. in 1975

End-use	Wool	Cotton	Non-cellulosics	Cellulosics
Apparel:				
U.S.A.	71.7	52.7	42.4	44.5
E.E.C.	69.7	43.2	55.4	37.3
Furnishings & other home uses:				
U.S.A.	23.3	29.5	33.0	23.9
E.E.C.	27.2	36.4	33.8	35.6
Industrial & other:				
U.S.A.	5.0	17.8	24.6	31.6
E.E.C.	3.1	20.4	10.8	30.1

Source: Based on Tables 1.6, 1.7, 1.12, 1.13, 1.18, 1.19, 1.23, and 1.24.

Studies by the International Wool Secretariat of total fibre consumption in wool-type end-uses (men's wear, women's wear, knitwear, carpets) indicate that in the major consumer countries (U.K., U.S.A., Germany, France, Italy, Japan, Holland and Belgium) total wool consumption declined from 495 m. kg. in 1967 to 448 m. kg. in 1977, that is a decline of about 10 per cent. In the same period, the proportion of wool to all fibres used in wool-type end-uses declined from 33 per cent to 16 per cent. This proportion is predicted by the International Wool Secretariat to fall further (to 11 per cent) by 1987 although total wool consumption might increase slightly. These trends are indicated in Fig. 1.18. An implication of this trend is that wool is likely to become increasingly a specialty fibre and in the view of J. Wilcox [29], North West Europe Area Director of I.W.S., will increasingly concentrate on the high price and quality section of the textile market.

Most of the world's wool is consumed in developed countries. It is interesting to consider the per capita level of wool and fibre consumption in the main developed areas of the world and compare this will figures for less developed countries. These statistics are set out in Fig. 1.19. One noticeable feature is that per capita wool consumption in the United States is much lower than in other developed areas. Per capita wool consumption in Western Europe and Japan is almost three times as great as in the United States. In Japan and Western Europe the proportion of wool consumed in relation to all fibres is considerably above the world average and is about five times the proportion in the United States (see Fig. 1.20). By world standards, Americans are relatively small users (per head) of wool.

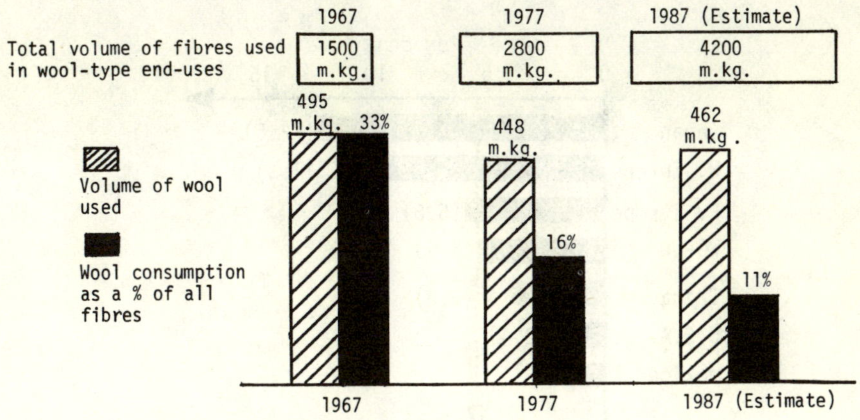

Fig. 1.18

Wool Consumption in Wool-Type End-Uses in Major Consumer Countries in m. kg. and as a Percentage of Total Fibres Used in These End-Uses 1967, 1977 and 1987

Source: Based on figures in "Why Wool?" [29].

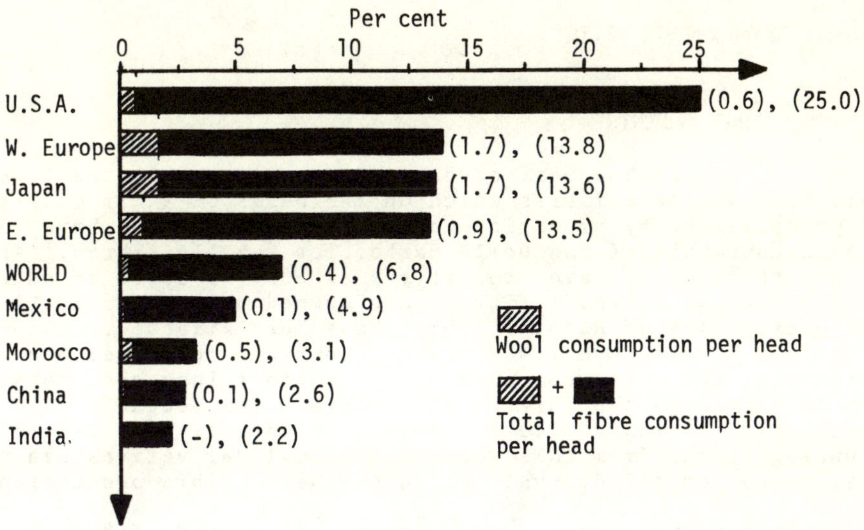

Fig. 1.19

Per Capita Consumption in kgs of Textile Fibres (with Per Capita Consumption of Wool Distinguished) for Selected Areas in 1974

Source: Based on statistics in "Wool's Position in the World Today - and Tomorrow" [4].

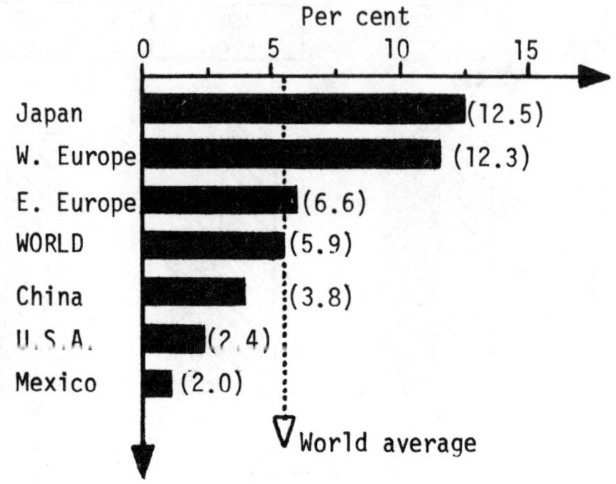

Fig. 1.20

Per Capita Wool Consumption as a Percentage of Total Per Capita Textile Fibre Consumption in Selected Areas 1974

Source: Based on Fig. 1.19.

1.5 CONCLUDING COMMENTS

Natural fibres have experienced great competition in the last 25 years from man-made fibres which on the basis of their qualities and price, aided by promotion and technical progress, have captured almost half of the world market for textile fibres. While not all these gains have been at the expense of sales of natural fibres, because man-made fibres have opened up new applications of fibres, sales of natural fibres have been affected. Even though oil prices have risen and may continue to do so, man-made fibres are likely to meet most of the world's increased consumption of fibres in the foreseeable future. Considerable oil price rises would be required to increase substantially the cost of producing fibres from this source, and coal derivatives can be substituted for oil derivatives in synthetic fibre production.

The invention and economic development of man-made fibres have done much to improve living standards in the world. The levels of per capita and total fibre consumption which have been achieved in the world in the last 25 years would have been impossible to achieve or would have been attained only at great sacrifice had man-made fibres not developed. Furthermore, competition from synthetic fibres has stimulated countervailing competition from natural fibres. Technological developments in cotton and wool textiles with new properties such as machine-washability and permanent-press are now available.

As a result of the growth of the man-made fibre industry, the nature of the fibre market has altered. The synthetic fibre industry, unlike the natural fibre industry, holds little stock of materials and its supply of raw materials is not subject to the vagaries of the weather. In contrast to the production of natural fibres, production of man-made fibres is in the hands of a few large companies located in the world's major industrial areas. Man-made fibre companies possess some market power and because of their fewness have incentives and possibilities to co-operate with one another. Man-made markets are oligopolistic and attention will be given to their oligopolistic nature in the next Chapter.

Fundamental changes have occurred in the textile fibre market in the last 25 years and these have posed new marketing problems for producers of natural fibres. The purpose of this book is to explore some of these problems especially in so far as they affect the marketing of wool and the instability of its price.

REFERENCES

[1] AIROV, J., *The Location of the Synthetic Fibre Industry,* Technology Press of Massachusetts Institute of Technology and John Wiley, New York, 1959.

[2] Australian Wool Corporation, *The Marketing of Australian Wool,* Melbourne, 1973.

[3] BONUS, J., "Man-Made Fibres and Natural Fibres: Competition in End-Uses in Western Europe", a paper presented at a conference, *The Future of Natural Fibres,* Shirley Institute, Manchester, November 1977.

[4] BREYER, F.W., and KESSELS, K.H., "Wool Place in the World Today - and Tomorrow", *Wool Record,* October, 33-35, 1977.

[5] Bureau of Agricultural Economics, *The Fibre Review 1974-75,* Australian Government Publishing Service, Canberra, 1976.

[6] Bureau of Agricultural Economics, *The Fibre Review 1975-76,* Australian Government Publishing Service, Canberra, 1977.

[7] Business Press Limited, "Cotton: dearer than it is worth?", *Textile Asia,* September, 18-19, 1976.

[8] COCKETT, S.R., *An Introduction to Man-Made Fibres,* Pitman, London, 1966.

[9] Comité International de la Rayonne et des Fibres Synthétiques (C.I.R.F.S.), *Information on Man-Made Fibres,* Paris, 1972 and 1973.

[10] Commonwealth Secretariat, *Wool Intelligence,* London, (various issues).

[11] Food and Agriculture Organisation of the United Nations (F.A.O.), *Agricultural Commodity Projections 1970-1980,* Rome, 1971.

[12] F.A.O. Commodities and Trades Division, *Impact of the energy Crisis on the Competitive Position of Agricultural Raw Materials,* Rome, 1974.

[13] HAGUE, D.C., *The Economics of Man-Made Fibres,* Duckworth, London, 1957.

[14] HIRSH, B.W., and ELLIS, P., *An Introduction to Textile Economics,* The Textile Trade Press, Manchester, 1974.

[15] HUFBAUER, G.C., *Synthetic Materials and the Theory of International Trade*, Duckworth, London, 1966.

[16] International Cotton Advisory Committee, *Cotton - World Statistics*, Washington, (various issues).

[17] International Federation of Cotton and Allied Textile Industries, *International Man-Made Fibre Statistics*, Zurich, (various issues).

[18] International Wool Textile Organisation and International Wool Study Group, *Wool Statistics*, London, (various issues).

[19] Japan Chemical Fibres Association, *Development Prospects of the Man-Made Fibre Industry in Asia and the Far East*, ECAFE Seminar, Japan, 1966.

[20] JONES, D.M., "The Present World Market for Polyester Fibres", *Textile Month*, July, 14-16, 1977.

[21] MARKHAM, J.W., *Competition in the Rayon Industry*, Harvard University Press, Cambridge, 1952.

[22] MONCRIEFF, R.W., *Man-Made Fibres*, Newnes-Butterworths, London, 1975.

[23] POSNER, M.V., "International Trade and Technical Change", *Oxford Economic Papers*, 13, 323-341, 1961.

[24] ROBSON, R., *The Man-Made Fibre Industry*, Macmillan, London, 1958.

[25] Textile Economics Bureau, *Textile Organon*, New York, (various issues).

[26] TRIFFIN, R.D., *Monopolistic Competition and General Equilibrium Theory*, Harvard University Press, Cambridge, 1940.

[27] United States National Advisory Committee on Food and Fibre, *Cotton and Other Fibre Problems and Policies in the United States*, Washington, D.C., July 1976.

[28] VERNON, R., "International Investment and Trade in the Product Cycle", *Quarterly Journal of Economics*, 80, 190-207, 1966.

[29] WILCOX, J., "Why Wool?", a paper presented at a conference on *The Future of Natural Fibres*, Shirley Institute, Manchester, November 1977.

CHAPTER 2

Marketing, Market Structure and Competition in the Fibre Industry

2.1 INTRODUCTION

There are striking differences between the structure of the man-made fibre industry and the structure of the cotton and wool industries. The former is dominated by a few large multinational companies whereas natural fibre production is in the hands of very many small farmers, that is small in comparison to the size of the man-made fibre producers. The development of man-made fibres has resulted in an increased concentration of market power in the fibre industry. This has come about partly as a result of the rise of large multinational firms producing man-made fibres which, in turn, has stimulated the development of economic organisations in the wool and cotton industries which, with government support, have intervened in the marketing of these fibres.

Multinational firms producing man-made fibres include E.I. Du Pont de Nemours, Imperial Chemical Industries, Hoescht, Akzo N.V., Asahi Chemicals and Courtaulds. The man-made fibre industry may be dominated by a few multinational firms because[1] (i) discoveries of new fibres have given particular companies temporary monopolies, for example, through the patent system, (ii) considerable cost reductions in the production of man-made fibres can be achieved from a large scale of production, (iii) extremely large amounts of capital are required to build polymer and fibre plants of an economic size, (iv) heavy expenditure on research and development (R & D) appears to be necessary if a company in the industry is to avoid economic stagnation and (v) considerable expenditure on the promotion of fibres, especially new fibres, is important to the success of the firm.

In most countries and economic regions, the man-made fibre industry is dominated by a handful of firms; it is oligopolistic. This market structure has given man-made fibres some advantage in their competition with natural fibres such as wool. For

---

[1] Data on the importance of these factors can be found in [1], [4], [13], [16], [24], [28] and [31].

TABLE 2.1

The Major Man-Made Fibre Producers of the World[(a)] Listed by Country and Ranked by Sales 1974 (Sales in Thousands U.S. Dollars)

Corporation	Country	Sales ($US mill)	World rank[(b)] as a textile producer by sales	World rank as a company by sales
Hoechst	West Germany	7821	1	23
E.N.I.	Italy	7172	2	26
I.C.I.	Great Britain	6912	3	28
Du Pont	U.S.A.	6910	4	29
Bayer	West Germany	6301	5	34
Montedison	Italy	6190	6	36
Goodyear	U.S.A.	5256	7	47
Phillips Petroleum	U.S.A.	4980	8	51
Eastman-Kodak	U.S.A.	4584	9	63
Rhône Poulenc	France	4234	10	73
Akzo	Netherlands	4009	11	78
Firestone	U.S.A.	3675	12	87
Monsanto	U.S.A.	3498	13	95
Courtaulds	Great Britain	2684	14	131
Gelenberg	West Germany	2640	15	137
Uniroyal	U.S.A.	2300	16	165
Allied Chemical	U.S.A.	2216	17	172
F.M.C. Fibers	U.S.A.	2074	18	180
Celanese	U.S.A.	1928	19	200
Asahi Chemical Industry	Japan	1811	20	210
American Cyanamide	U.S.A.	1780	21	218
Kanebo	Japan	1638	22	241
Hercules	U.S.A.	1525	23	264
Toray Industries	Japan	1433	24	276
Toyobo	Japan	1404	25	285
Teijin	Japan	1079	26	367
Coats Patons	Great Britain	1048	27	374
Rohm and Haas	U.S.A.	1022	28	386
Snia Viscosa	Italy	831	29	474
Owens Corning Fiberglass	U.S.A.	828	30	475
Unitika	Japan	800	31	484
Mitsubishi Rayon	Japan	778	32	497

Notes: (a) In terms of sales.

(b) In terms of *all* sales not only textile sales.

Source: Based on data in *Fortune* [39].

instance, prices of man-made fibres have been fairly stable compared with those for wool and in recent years, cotton. Most textile manufacturers consider price stability to be an advantage.

Sellers in both wool and cotton markets have attempted to organise to meet competition from the man-made fibre industry. In the case of wool, the International Wool Secretariat, representing Australia, New Zealand, South Africa and Uruguay, promotes wool in Europe and other selected areas on behalf of those participating countries. The activities of the International Wool Secretariat are in general concerned with wool textile marketing and it seeks to promote the use of wool by textile producers and consumers. In addition, Australia, New Zealand and South Africa, major wool producers, have established statutory boards (in Australia, the Australian Wool Corporation), to regulate the marketing of raw wool. Cotton production is also regulated in a number of countries and the International Institute for Cotton undertakes similar activities to the International Wool Secretariat. It promotes the use of cotton in Europe and Japan on behalf of its thirteen cotton-producing member countries.

Let us examine the market structure of the man-made fibre industry in more detail, concentrating on the situation in the United States and Japan and then discuss the competitive strategies of manufacturers of man-made fibres before dealing with the marketing of cotton and wool.

2.2 MARKET STRUCTURE OF THE MAN-MADE FIBRE INDUSTRY

The manufacture of man-made fibres is dominated by large multinational companies. If one examines the 500 largest companies in the world ranked by sales, at least thirty-two of these produce man-made fibres. These are listed in Table 2.1. This Table reveals that most companies in the group are American or Japanese but that on the whole Japanese companies are smaller than their European and American counterparts. Many companies in the group are substantial chemical or petrochemical producers. These include Hoescht, I.C.I., Du Pont and Bayer. Most are diversified producers - have economic interests in fields other than the man-made fibre industry. Nearly all of these firms are engaged in multinational operations.

Consider the pattern of man-made fibre production in the United States by companies. This is indicated in Table 2.2.

Table 2.2 reveals that all the main United States producers, with the exception of Courtaulds N.A., derived a substantial proportion of their sales receipts from other than fibres. In fact only two companies in this group (Courtaulds excepted) had fibres as their main source of sales receipts and this source did not exceed much more than half.

Excluding companies producing solely glass fibre or solely olefin, there are about 14 fibre producers in the United States. However

TABLE 2.2

Sales of United States Fibre Producers 1973 and 1974 (m.$US)

Company	Corporate total sales		Approximate fibre sales		Percentage fibre to total corporate sales		Fibre types produced[a]
	1973	1974	1973	1974	1973	1974	
Du Pont	5275	6910	2057	2418	39.0	35.0	A, Ac, E, F, Pa, Pe
Celanese	1609	1928	1036	1123	64.4	58.2	Ac, Pa, Pe, T
Monsanto	2647	3998	695	788	26.3	22.5	A, M, Pa, Pe
Akzona	701	753	385	380	54.0	50.4	Pa, Pe, R
Eastman	4035	4583	305	349	7.6	7.6	Ac, M, O, Pe
FMC	1719	2074	295	300	17.2	14.5	Ac, Pe, R, V
Allied Chemical	1664	2216	226	289	13.6	13.0	Pa, Pe
Dow-Badische	282	380	140	165	49.6	48.5	A, Pa, Pe
Phillips	2990	4980	151	161	5.1	3.2	O, Pe
American Hoechst	339	386	142	140	41.9	36.3	Pe
Rohm & Haas	789	1021	103	122	13.0	12.0	Pa, Pe
El Paso (Beaunit)	976	1258	108	105	10.9	8.4	O, Pa, Pe, R
American Cyanamid	1472	1780	65	80	4.4	4.5	A
Courtaulds, N.A.	60	60	60	60	100.0	100.0	Pa, R

Note: (a) Fibre types: A - acrylic, Ac - acetate, E - elastane, F - fluorocarbon, M - modacrylic, Pa - nylon, Pe - polyester, R - rayon, T - triacetate, V - vinyon.

Source: Textile Month [34].

sales are concentrated. The top four companies in terms of fibre sales accounted for 73 per cent of sales by these companies in 1974. Du Pont, Celanese and Monsanto dominated the industry. Fig. 2.1 indicates the percentage of total fibre sales accounted for by the main producers of textile fibres in the United States.

If we consider individual types of man-made fibres, there are few producers of some, for instance, acrylics. Table 2.3 lists the number of U.S. producers of each fibre-type in 1975. There are a considerable number of producers of polyester and nylon fibre but often these fibres are considerably differentiated and different varieties of the same type can be very poor substitutes. Product differentiation tends to reduce the extent of competition in the man-made fibre industry.

Dr Leland S. Liang has observed that:

> From a handful of man-made fibre producers fifteen years or so ago, there are now some two dozen firms in the industry. To be sure the industry is still dominated by three or four fibre producers [20, p. 1].

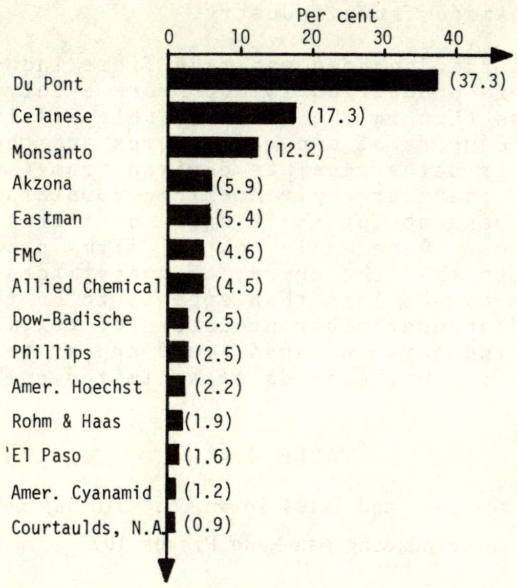

Fig. 2.1

Fibre Sales of Industrial United States Textile Producers as a Percentage of Their Aggregate Fibre Sales 1974

Source: Based on Table 2.2.

TABLE 2.3

Number of United States Producers of Selected Fibre Types 1975

Fibre type	Number of producers
Acetate and Triacetate	3
Acrylic	5
Aramid	1
Modacrylic	3
Nylon	25
Polyester	21
Rayon (Viscose)	5
Cuprammonium Rayon	1

Source: *1975 Man-Made Fibre Yearbook* [29].

But as will be discussed later, firms have been forced to alter their marketing strategies by the growth in the number of producers in the man-made fibre industry.

An examination of the Japanese man-made fibre industry indicates that man-made fibre production is much more evenly spread amongst the main producers than in the U.S.A. Table 2.4 lists the 11 major Japanese producers of man-made fibres and the approximate percentage of their sales receipts derived from this source tends to be much higher than for their American counterparts. These firms are more dependent for their sales on fibres than are United States firms. Once again not all firms produce every fibre type and variety so that the number of competitors supplying particular fibres may be less than appears to be the case [29]. If one takes into account minor producers of textile fibres, the total number of producers in Japan would appear to be 15 or less [19]. Production is in the hands of a limited number of companies.

TABLE 2.4

Number of Employees and Sales in ¥m. of Main Japanese Firms Producing Man-Made Fibres 1977

Name of company	Number of employees	Sales ¥m	Man-made fibre sales ¥m	Man-made fibres as a % of total sales
Toray Industries	17839	418982	263959	63.0
Asahi Chemical Industry	16316	473430	241449	51.0
Teijin	11568	349521	241169	69.0
Toyobo	18108	280347	171012	61.0
Kanebo	14312	415977	141432	34.0
Mitsuibishi Rayon	4867	144977	140383	72.0
Kuraray	6888	175324	133246	76.0
Unitika	11128	245389	125148	51.0
Toho Rayon	2529	64552	50351	78.0
Nisshin Spinning	6548	111317	48979	44.0
Fuji Spinning	5330	73592	36060	49.0

Source: Based on data in *Japan Economic Yearbook 1977-78* [38].

The percentage of the aggregate fibre sales of the 11 Japanese textile firms mentioned in Table 2.4 accounted for by each is shown in Fig. 2.2. Comparing this Figure with Fig. 2.1, we see that market concentration in Japanese synthetic fibre production is less than in the United States.

As has been the practice of American and European producers of man-made fibres, Japanese fibre producers have established plants abroad. Plants have been established in South Korea, Taiwan, Thailand, Malaysia, Phillipines, Indonesia, Brazil, Venezuela and Spain. Most of these have been established in the last decade as Japan has attempted to increase foreign investment and make provision for economic nationalism in other countries and the slowdown in the growth of its domestic textile industry.[2]

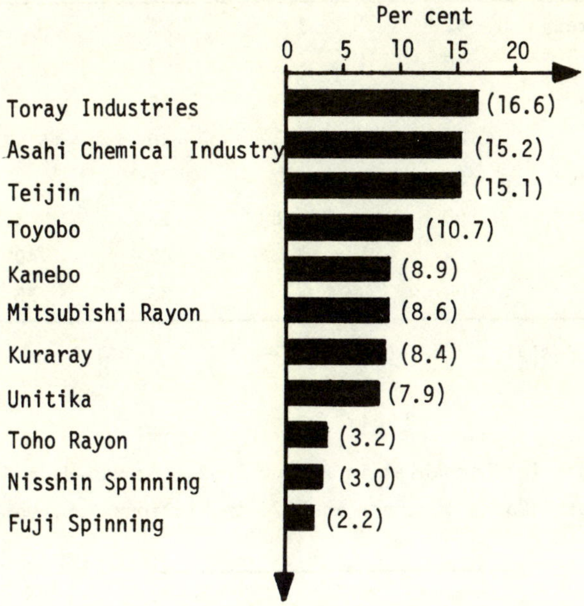

Fig. 2.2

Fibre Sales of Individual Japanese Textile Producers as a Percentage of Aggregate Fibre Sales 1977

Source: Based on Table 2.4.

In some South East Asian countries, the amount of textile capacity in the hands of Japanese controlled firms is considerable The shares involved are set out in Table 2.5.

Within the European Economic Community, the main producers of man-made fibres are Hoescht, I.C.I., Courtaulds, E.N.I., Bayer, Montedison, Rhône Poulenc, Akzo, Gelenberg, Coats Patons and Snia Viscosa. Subsidiaries of American and Japanese firms also produce in this area but the number of large producers possibly does not exceed 16.

---

2 For a discussion of the trend see [42].

TABLE 2.5

Share of Textile Capacity in South East Asia Controlled by Japanese Controlled Firms in Mid-1975 in Per Cent

Country	Synthetic fibres	Synthetic yarn	Spinning	Cloth
South Korea	42	1	3	-
Taiwan	10	-	3	-
Hong Kong	-	40	5	-
Thailand	85	12	40	16 - 20
Phillipines	32	-	2	-
Indonesia	100	7 - 8	15	1.5
Singapore	-	13	47	40
Malaysia	100	-	33	33

Source: *Textile Asia* [9].

TABLE 2.6

Selected Industrial Companies in Europe Producing Man-Made Fibres Ranked by Total Sales Turnover in 1976 in Millions U.S. Dollars

Company	Turnover
Hoechst	9941.24
Bayer	9802.36
I.C.I.	7039.42
Montedison	5296.50
Akzo	4375.16
Rhône Poulenc	4373.28
Courtaulds	2571.13
Du Pont	1083.00
Snia Viscosa	893.27
Monsanto Europe	697.00

Ten of the most important European producers of man-made fibres are ranked in Table 2.6 by their total value of sales in 1976. However a larger proportion of their sales is earned from other than textiles. It is reported that Hoechst and I.C.I. earn only

10 per cent of their turnover from synthetic fibre production, and for Bayer A G the proportion amounts to six per cent [36]. The percentage of total sales earned from synthetic fibres by Rhône-Poulenc has fallen and is now around 27 per cent. *Textile Month* [36, p. 23] reports that "perhaps of all the major European groups only Akzo rely more heavily on synthetic fibre production for a greater percentage of their turnover" than Rhône-Poulenc. Hence it is clear that most European producers of man-made fibres are less reliant on these fibres for sales than their Japanese and American counterparts (compare Tables 2.4 and 2.2). The contrast between the Japanese and European positions is most marked.

Taking America, Europe and Japan together there appear to be about 30 independent sizeable producers of man-made textile fibres and between them and their subsidiaries they control practically all man-made fibre production in non-communist countries.

Economies of scale and capital requirements for efficient manufacture, and promotional and research barriers may help to explain the relative fewness of producers. But since World War II the number of producers has expanded. Petrochemical producers have entered the industry in America, Europe and Japan and the patents on some important fibres have expired or are about to expire, for instance on nylon. The international cartel-like structure of the industry which existed before World War II appears to have disappeared although it is still possible even if more difficult, for the main producers to co-operate in dealing with their fibre marketing problems.

2.3 COMPETITIVE STRATEGIES OF MAKERS OF MAN-MADE FIBRES

Competition between fibres and fibre producers occurs in many different ways. Factors which are important in competition include the relative price of a fibre, its qualities, advertising and promotion of the fibre, technical advice to textile producers using the fibre and in some instances the extension of credit to processors of the fibre.

Factors contributing to the expanding sales of man-made fibres have been (i) their lower price, (ii) their superior qualities in some end-uses, e.g., the easy-care qualities of slacks made from some man-made fibres, (iii) their superior qualities in the production of some textiles, for instance, synthetic yarns are less likely to break when knitted at high speed, (iv) the relative stability of their prices, (v) the contributions (subsidies) of makers of man-made fibres to the advertising of processors and retailers using these fibres - subsidies of up to 50 per cent have been common, (vi) the provision of promotional material by makers of synthetic fibres to commercial concerns producing and merchandising goods containing these fibres, for instance, the provision of display material, pattern books and so on, (vii) synthetic fibre makers generously provide technical services to processors using their fibres - this "back-up" service forms a major part of the selling effort of makers of synthetic fibres and (viii) sometimes makers of man-made fibres extend longer than normal credit terms to their customers and provide them with loans for the installation of machinery.

Makers of man-made fibres spend heavily on research and development. As patents on existing fibres expire, great effort is made by existing patentees (owners of these patents) to discover superior synthetic fibres so as to enable them to hold their market position more easily. For instance, Du Pont after its discovery of nylon increased its efforts to find superior types of nylon and substitutes for nylon such as polyester.

The man-made fibre industry illustrates Joseph Schumpeter's thesis that in capitalism the competition which counts is:

> ... the competition for the new commodity, the new technology, the new source of supply, the new type of organization (the largest-scale unit of control for instance) - competition which commands a decisive cost or quality advantage and which strikes not at the margins of the profits and the output of existing firms but at their foundations and their very lives [32, p. 84].

Indeed, Markham [23] suggests that Schumpeter drew much of his inspiration for his thesis from his observations of the man-made fibre industry.

The importance of advertising and promotion in helping the market penetration of man-made fibres has been widely recognised. For instance, the U.S. National Advisory Committee on Food and Fibre reported in 1967:

> Quality characteristics of the noncellulosic fibres in combination with large-scale promotion and advertising appear to be largely responsible for the sharp rise in their use. The strategy of the producers of the man-made fibres has been outlined by Berrer [3]. First, intensive research identifies potential markets, and then the fibre is tailored for the end-use. Second, a marketing plan is developed which may include the following items (i) technical assistance at all stages of manufacturing, (ii) a quality control system to protect the trade mark of the fibre maker on ready-made goods, (iii) securing of good profit opportunities for co-operating manufacturers, and (iv) effective advertising and public relations [24, p. 34].

Research and promotion expenditure in the man-made fibre industry in the 1960s far exceeded the expenditure in support of natural fibres. For example, the U.S. National Cotton Council estimated that in 1965 manufacturers of man-made fibres spent $US135 million on research and $US70 million on promotion whereas the corresponding expenditures for cotton were only $US26.5 million and $US4 million respectively. However, since the 1960s cotton has increased research and promotional effort.

Because most companies producing synthetic fibres produce other products and because they do not give a breakdown as a rule of the composition of their R & D expenditure, it is difficult to obtain accurate estimates of their R & D expenditure on synthetics. Table 2.7 gives the overall intensity of research (R & D expenditure to sales) by five large European industrial companies which also produce man-made fibres. R & D expenditure as a proportion of their total sales in 1976 was 3.6 per cent. This is a relatively high figure but its size may reflect the higher research intensities of some of the other productive activities, e.g. chemicals, undertaken by these companies. The R & D expenditure of Montedison on synthetics is available. This indicates an R & D intensity for synthetic fibres of about 1.7 per cent. Taking all companies together R & D intensity for synthetics is probably about two per cent of sales. When this is combined with a likely minimum of expenditure on advertising of fibres of two per cent of sales, research and promotion expenditure in the man-made fibre industry appears to account for approximately four to five per cent of sales turnover.

TABLE 2.7

Total Research and Development Spending as a Proportion of the Total Corporate Sales of Five Companies (Akzo, Bayer, I.C.I., Montedison, Rhône-Poulenc), Total R & D Spending of Montedison as a Proportion of its Total Sales, and R & D Expenditure on Synthetics by Montedison as a Proportion of its Sales of Synthetics

Item	1971	1972	1973	1974	1975	1976
R & D as a % of total sales of the five companies	3.6	3.8	3.4	3.1	3.3	3.6
Total R & D of Montedison as a % of its sales	2.6	2.5	1.8	1.4	2.0	1.9
R & D expenditure on synthetics by Montedison as a % of its sales of synthetics	2.5	1.7	1.3	1.5	1.7	1.7

Source: Based on material in "Wool: Implications of Development in the Synthetic Fibre Industry" [45].

The importance of promotion to the success of new fibres is underlined by Dr Hermann Zwick, Director of Hoechst, in his comments on his company's successful introduction of Trevira, a polyester fibre. Considerable market research proceeded the introduction of the fibre and Dr Zwick concluded from his company's experience that:

> ... the most important elements in the marketing rise during the introductory phase were an intensive national advertising campaign aimed at the general public. Sales promotion at all levels in the textile pipe-

> line as well as the building up of effective sales organization and Technical Development and Service Department [48, p. 51].

Price has also played a significant role in interfibre competition but direct or overt price competition between manufacturers of man-made fibres has been rare. Price leadership seems to be common and Du Pont is reputed to be the price leader in the United States for many fibres [21], [28]. In general, the real prices of new synthetic fibres have tended to decline by steps, remaining *stable* for various intervals and declining discretely.

Albach [2], [3] found in his study of the pricing of polyamides in Germany that indirect price competition is much more important than overt price competition in the short-run. Quoted prices for man-made fibres (which are relatively stable) often differ from effective prices. The invoice price of a fibre is not necessarily its effective price because producers give bonuses, aids and credits (sometimes secretly) to buyers:

> In addition to annual bonuses [paid to buyers on the basis of their total purchases] and other rebates, suppliers grant credit aids, financial help and advertising subsidies which cannot be easily allocated to an individual transaction or to individual products. However, they are a token of the close business relationship between the supplier firm and its customer. It seems fair to say that the intensity of competition in oligopolistic markets can be measured by the importance of such aids and subsidies in doing business [3, p. 96].

In Albach's view these aids are parameters of indirect price competition, are essential parts of pricing policy on the side of suppliers and measure the intensity of competition in this oligopolistic market. Stability of quoted prices in the man-made industry is not adequate evidence of lack of price competition.

While man-made fibre producers use methods of indirect pricing to avoid open price reductions and to strengthen their goodwill with customers, these methods "are also used to persuade customers of competing firms to switch from one supplier to the other without triggering off a price war among the oligopolistic suppliers in the market" [3, p. 96]. In the long-term, however, quoted prices alter to reflect underlying conditions.

It was reported in 1967 that:

> ... according to a spokesman for Du Pont, price reductions for non-cellulosic fibres reflect many factors. First of all raw material costs have been declining. Second, efficiency of production has been increasing, more than one yarn can now be spun from a spinneret and larger yarn bundles can be converted to staple or spun yarn. Third, there have been economies of large scale production [24, p. 12].

However, it is no longer true since the energy crisis that raw material costs are falling in the man-made fibre industry. As Table 2.8 indicates the price of some raw materials required for the production of synthetic fibres has doubled or more than doubled since the oil crisis of 1973. However since raw material costs are a small component of the total costs of production of synthetic fibres, this has not meant as large a rise in total cost of production. For instance, it has been estimated [22] that the cost of polyester production rose by about 20 per cent as a result of the oil crisis.

TABLE 2.8

Index of Synthetic Fibre Raw Material Prices Japan 1970-1975 (1970 = 100)

Synthetic fibre raw material	1970	1971	1972	1973	1974	1975
Polyester:						
D.M.T.[(a)]	100	71	71	73	95	114
Ethylene Glycol	100	99	99	103	159	204
Acrylic:						
Acrylonitrile	100	95	91	92	149	183

Note: (a) D.M.T. - Dimethyl Terephthalate.

Source: "Implications of Developments in the Synthetic Fibre Industry" [45, p. 120].

Price leadership which had been mentioned by Markham [23] in the 1950s as being important in the man-made fibre industry still appears to be important. Mr Claude Ramsey, Chairman and President, Akzona Incorporated, which has American Enka as a subsidiary, said in 1974:

> America Enka is a medium-sized fibre producer - the fourth largest in the U.S. ... We must adapt our strategy to our relatively smaller size. Normally, we try to out-innovate those producers above us, to outmanoeuvre the three producers below us, and to avoid head-on price competition with the larger producers of commodity fibres [28, p. 58].

Mr Ramsey goes on to point out that:

> While commodity fibres sold on the brutally competitive basis of quality, price and service form the framework of a producer's business - certainly they represent 75 per cent of ours - it is the specialty fibres, with no more than one competitor, which account for the profitable growth of the company.
>
> Specialty fibres carry a premium, convoy commodity fibres in blends, cement customer relations - who is so big that he can afford *not* to do business with an innovator? and finally, properly advertised, help our customers move merchandise in volume at a profit [28, p. 58].

All manufacturers of man-made fibres recognise the importance for their growth and profits of developing new fibres and specialty products and promoting them. Advertising intensities are still high in the industry but not as high as in the 1960s when according to Mr Ramsey fibre advertising by man-made fibre producers represented an estimated *minimum* of two per cent of sales, an intensity much higher than for natural fibres. Mr Ramsey explained that:

> ... emerging producers from the petrochemical sector poured on the advertising dollars to gain position. Established producers were launching innovative fibres for hosiery, carpets and so on. Some expanded their product mix to include the polyesters. All jockeyed for a position in the fibre trademark sweepstakes, putting heavy advertising investments on the line [28, p. 61].

At the end of the 1960s however, the level of direct advertising support to manufacturers of softgoods and retailers was reduced.

Competitive strategies employed by the man-made fibre industry reflect its oligopolistic structure. Price stability, price leadership, research effort and marketing emphasis on advertising and promotion are characteristic of most oligopolistic industries.

Despite the oligopolistic nature of the man-made fibre industry, it has not been able to avoid economic difficulties in the current world recession. Due to considerable investment in new capacity, especially in Europe, the industry is suffering from considerable excess capacity. *Textile Month* [35] recently reported that "the overall utilisation of productive capacity in the E.E.C. reached no more than 72%, and it is acknowledged that last year the gap widened a significant amount further". It is estimated that by 1985 utilisation may be no more than 60 per cent since capacity is expected to expand at a faster rate than consumption. This has meant that prices of synthetic fibres have remained depressed and that large financial losses have been incurred by manufacturers of man-made fibres. While imports to the E.E.C. have been blamed for some of the difficulties in the area and a new Multi Fibre Arrangement is being negotiated to cope with these, the main problem seems to be an internal one of over-investment in new synthetic fibre capacity. Already some of the largest synthetic fibre producers are rationalising to reduce capacity [35], [36].

2.4 COTTON - MARKETING AND COMPETITIVE STRATEGIES OF THE COTTON INDUSTRY

Many of the countries producing cotton have adopted measures to influence the production, price and marketing of raw cotton. Price supports and guarantees and subsidies on inputs have been common, and expansion of production has been encouraged in developing countries and centrally planned economies. At the present time, the possibility of an international buffer stock scheme for cotton is being discussed by the United Nations [41].

Until recently the United States was the world's major producer of cotton (its production is now exceeded by that of the U.S.S.R., and Chinese production is nearly as great) and still is the major exporting country by a wide margin. At least until recently, world cotton prices were determined largely by United States cotton policy.

Prior to 1965, attempts by the United States Government to support cotton prices led to large stocks of cotton being accumulated by the Commodity Credit Corporation of the United States and had a stabilising influence on world cotton prices. However, the policy seemed to be concerned more with income redistribution considerations rather than meeting competition from man-made fibres. Indeed, Hakim concludes in his Ph.D. dissertation that U.S. cotton policy by maintaining the international price of U.S. cotton in this period encouraged market penetration by man-made fibres and growth of cotton production in foreign countries [14].

However, after the introduction of the *Agricultural Act of 1965*, guaranteed cotton prices were reduced to a level sufficiently low to have no influence on market prices, C.C.C. stocks were rundown and payments to farmers for acreage diverted out of cotton were introduced. United States cotton production began to decline and has continued to do so.

At present the U.S. upland cotton crop is governed by the *Agricultural and Consumer Protection Act of 1973* [43]. The major provisions of the programme for the 1975 cotton crop were (i) a guaranteed target price of 38 cents per pound, (ii) a preliminary loan rate (advanced against the crop) of 90 per cent of the average world price for cotton, (iii) no requirement to limit acreage to qualify for support.

At present the guaranteed United States cotton price seems to be slightly below world cotton prices and broadly U.S. policy is one of keeping U.S. cotton prices roughly in line with world prices and avoiding a build-up of cotton stocks. This has been reflected in falling levels of cotton production in the U.S., and the greater instability of world cotton prices from 1967-68 onwards.

The cotton industry has been much slower than the wool industry in counteracting the research and promotion efforts of the man-made fibre industry. It was not until after 1966 that any significant collective action emerged. The United States *Research and Promotion Act of 1966* authorised the imposition of a levy on cotton growers to support cotton research and promotion. In 1975 the levy amounted to $US1.00 per bale and $US14m. was authorised for expenditure from this source. In addition the *Agricultural Act of 1970* as amended by the *Agriculture and Consumer Protection Act of 1973*, provides for financial support for cotton research. However, despite increased research into and promotion of cotton in the last decade, expenditure is small compared to that by manufacturers of man-made fibres.

In the United States the National Cotton Council is responsible for the advertising and promotion of cotton. The International

Institute for Cotton, founded in 1966, is responsible for the promotion of cotton in Western Europe and Japan. The Institute is financed by the governments of its member countries which include Brazil, Greece, India, Mexico, Peru, Spain, Tanzania, Uganda and the United States. Financial contributions are based on exports of spinnable raw cotton to Western Europe and Japan.

The budget of the International Institute for Cotton is not large. It is less than that of the International Wool Secretariat and the promotional intensity of its effort is smaller.

A second international body, the International Cotton Advisory Committee, with members from cotton consuming and producing countries, disseminates market information about cotton and provides a direct liasion between growers and users.

A recent report by the Economic and Market Research Department of the National Cotton Council of America concluded:

> Research, along with promotion and price remains the cornerstone of a sound competitive industry.
>
> Research and its implementation are vital and continuing needs in keeping cotton production competitive with other fibres and other crops. It is likely that a combination of breakthroughs in many different areas will be required to get the job done. Critical areas relating to consumption are cotton dust, flame resistance and durable press for light weight garments. The increasing costs of many production inputs are energy-related and research is needed to cut requirements in this area [8, p. 110].

The report goes on to argue that strong promotional programmes to maintain and improve cotton's share are essential and that present efforts may be on too small a scale:

> In spite of stepped-up producer funding of research and promotion in the last decade, the cotton industry, its suppliers and government combined still spend less on these vital activities than the man-made fibre industry. Increased research and market development funding from all possible sources seems to be the requirement for cotton to compete effectively in the years ahead [8, p. 110].

On the pricing side, the report points out that the competitive position of cotton will depend on its future price in relation to the prices of man-made fibres and reasonable price stability of cotton may be needed if cotton is to retain its market share.

The low relative price of man-made fibres is currently helping to depress the demand for cotton. In the short-term this position is not expected to change significantly because of the amount of excess capacity in man-made fibre production. Indeed:

> ... a further increase in global MMF production capacity is not anticipated in the near term although one is planned in Japan and significant plant retirements are possible within the EEC. With rates of capacity utilisation remaining below normal, further (although probably limited) aggregate capacity growth is likely to continue to restrain the extent of any price rises realised by MMF producers. Major improvements in MMF prices appear to be unlikely therefore in the shorter term, with increases in

> many industrialised countries importantly dependent upon progress of rationalisation measures and on consumer confidence and demand [7, p. 243].

The Australian Bureau of Agricultural Economics predicts that:

> In the longer term *world demand* for cotton should expand gradually as growth in global population and economic activity proceeds. Future trends in cotton prices will depend on this demand growth, changes in supplies and demand shifts arising from man-made fibre competition. That competition should ease in the medium term in association with the anticipated rationalisation of the man-made fibre industry which, other things equal, is expected to result in increased man-made fibre prices. Over the longer term, technological advances in man-made fibre production will provide continuing competition for cotton [7, p. 245].

2.5 WOOL MARKETING AND COMPETITIVE STRATEGIES OF THE WOOL INDUSTRY

The wool industry has been relatively more active than the cotton industry in promoting its fibre and counteracting the competitive activity of makers of man-made fibres.

A first step towards co-ordinated promotion of wool by major wool growing and exporting countries was taken in 1937 with the establishment of the International Wool Secretariat by Australia, New Zealand and South Africa. Today these countries plus Uruguay form the membership of the International Wool Secretariat, an organisation with offices in the main wool consuming areas of Europe, Japan and North America and is intended to promote the use of wool by advertising and promotion.

Operation of the Secretariat were at first on a modest scale but it became clear during the 1950s that manufacturers of synthetic fibres were spending much more than the wool industry on research and development, technical service to processors and advertising campaigns. The importance of giving technical advice to textile and clothing manufacturers using wool and the need to develop and adequately evaluate and transfer wool textile inventions for industrial application became apparent. In the absence of sufficient developmental facilities and pilot plants in wool consuming countries, it had proven difficult to transfer wool textile inventions from wool producing countries to textile producers in wool consuming countries. Accordingly a five year plan was introduced in 1962 to expand the activity of the International Wool Secretariat. Technical and research centres were expanded and increased in number and pilot plants established in nine countries to demonstrate new techniques. Promotional activity was expanded and an International Wool Fashion office was established in Paris.

Woolmark was launched in 1964 to serve as an international symbol for pure wool, to identify products made from pure wool and establish a quality image for them. In moving in this direction, the International Wool Secretariat adopted similar policies to the brand-name policies used by makers of man-made fibres. In 1971, Woolblendmark was also introduced by the I.W.S. to identify wool-rich blends [18].

In 1974-75 and again in 1975-76 the International Wool Secretariat spent almost $A45 million in promoting wool in one way or the other. Direct marketing expenditure amounted to about three-quarters of this sum and spending on technical development and advice was around 10 per cent of the sum. Contributions to the budget of the Secretariat are based upon the wool exports of its member countries. Australia provides a little less than two-thirds of the income of the Secretariat, New Zealand a quarter and South Africa about 10 per cent. Uruguay makes a small cash contribution. The contribution of the various countries are financed partly by government subsidies and partly by levies on wool growers.

The International Wool Secretariat is controlled by a board of 14 members. Membership and financial contributions do not run exactly parallel. Relative to their financial contribution, New Zealand and Australia are under-represented on the board and South Africa is over-represented. Half the board of 14 are from Australia, three each from New Zealand and South Africa and one from Uruguay. The relative importance of the various types of promotional activities of the International Wool Secretariat is indicated in Table 2.9.

TABLE 2.9

Expenditure by the International Wool Secretariat in 1974-75 Classified According to Purpose

Type of activity	$A million	Per cent
Technical:		
Research & Development	2.4	5.9
Service to industry	2.1	5.1
Marketing:		
Media advertising:		
institutional[(a)]	12.1	29.3
joint promotions[(b)]	9.2	22.3
Salaries & travel	5.2	12.7
Services & assistance to industry	3.7	8.9
Licencing & quality control	1.8	4.3
General management & finance	4.8	11.5
Total	41.3	100.0

Notes: (a) Institutional advertising promotes all wool and is epitomised by the Woolmark programme.

(b) Joint promotions are funded by the I.W.S. and manufacturers together for specific wool products.

Source: *Draft Report on Financing Promotion of Rural Products* [17, pp. 7, 8].

It is interesting to compare the International Wool Secretariat's expenditure on institutional promotion with its expenditure on joint promotion. The latter is almost as large as the I.W.S.'s expenditure on institutional promotion and gains greater significance when one considers that in addition to $A9.2 million contributed by the I.W.S. for joint promotion campaigns, the industry and trade contributed $A15.4 million. In theory, joint promotion should enable the International Wool Secretariat's limited marketing budget to be used more effectively.

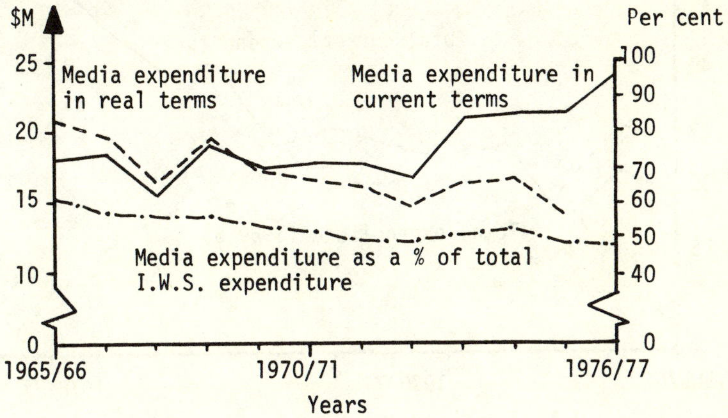

Fig. 2.3

I.W.S. Expenditure on Media Advertising in Current and Real Terms and as a Percentage of Total I.W.S. Expenditure 1965-66 to 1976-77

Source: Based on Table 3 in *Draft Report on Financing Promotion of Rural Products* [17, p. 9].

Although the level of expenditure on wool promotion was significantly raised in the earlier 1960s, the real level of this expenditure has not been maintained. The level of expenditure by the I.W.S. on media advertising is graphed in Fig. 2.3. It can be seen from the Figure that in real terms expenditure on media advertising has fallen by about 25 per cent since 1965-66. Expenditure on media advertising has also fallen as a percentage of I.W.S. total expenditure.

Again, while monetary expenditure by the International Wool Secretariat has risen steadily since 1965-66 it has failed to keep pace with the rate of inflation and as illustrated in Fig. 2.4, slowly declined in real terms. Indeed, since 1968-69 real expenditure by the I.W.S. has declined by about 30 per cent. In an article discussing the problem posed for the I.W.S. by the fall in its spending in real terms, *Wool Record* reported that:

> Budget increases had so far not kept pace with inflation. The I.W.S. had closed branches, grouped their offices and made many economies. But Dr Laxer [managing director of the I.W.S.] said: 'We can't every season ask for an increase in budget to compensate for our loss of purchasing

power'. He suggested that relatively more might have to be spent on 'men on the ground' and less on advertising [10, p. 16].

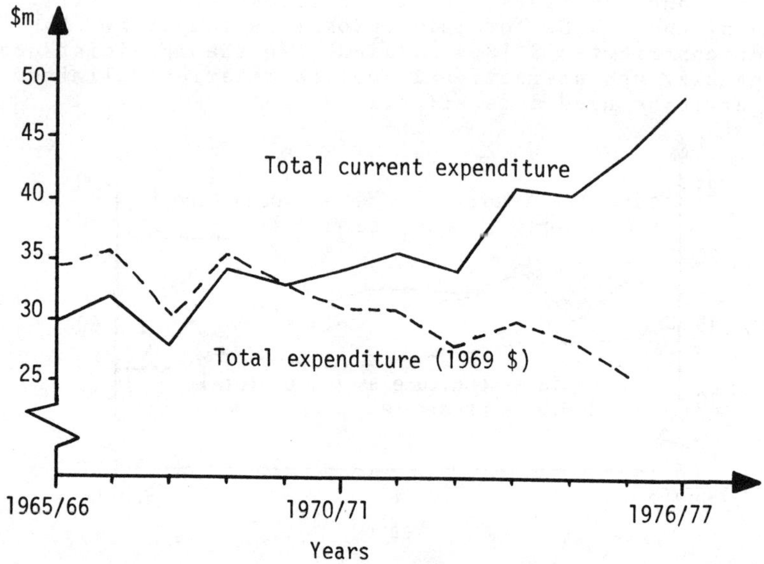

Fig. 2.4

Total I.W.S. Expenditure in Current and Real Terms (1969 dollars) 1965-66 to 1976-77[(a)]

Note: (a) 1976-77 figures are budgeted.

Source: Based on Table 3 in *Draft Report on Financing Promotion of Rural Products* [17, p. 9].

It might also be noted that the Australian Wool Corporation is directly responsible for the promotion of wool in Australia. In the year ended 30 June 1976 it spent about $A3 million for this purpose and the Corporation opened a U.S. office in 1975 to help promote the sale of Australian wool to the United States.

In addition to the support which wool producing countries gave to the International Wool Secretariat, each conducted its own research into wool. In recent years Australia has spent about $A15 million annually for this purpose. The lion's share of these Wool Industry Research Funds is allocated to the Commonwealth Scientific and Industrial Research Organization which has been able to make important scientific advances on behalf of the industry.

The efforts which the major wool producing countries are making to improve the market for wool is underlined when one considers that in 1975-76 more than three per cent (3.3) of the value of the Australian wool clip was used for the promotion of wool and almost two per cent for research. These intensities may not be substantially different from those of many man-made fibre manufacturers at present. New Zealand spent about three per cent of the value of its clip on promotion and research and South Africa also spent a substantial amount in this way.[3]

While a promotion intensity of three per cent of sales may seem high, it may be smaller than that of leading manufacturers of man-made fibres and of course this intensity only applies to a part of total world wool sales. In Australia, New Zealand and South Africa, governments subsidise the research or promotion efforts of the wool industry to some extent. The promotion efforts and strategies of the wool industry will be discussed in depth in Chapter 9.

Instability of wool prices has been a continuing problem for the industry. It has caused the incomes of woolgrowers to fluctuate considerably and has possibly reduced the demands of textile producers for wool. The Australian Wool Corporation in its report, *The Marketing of Australian Wool*, said:

> It is not possible to provide empirical evidence of the extent to which the demand for wool is disadvantaged by fluctuating wool prices. It can be recorded, however, that wool textile spokesmen have frequently stressed the serious difficulties created for them by price instability... A recent example was the public statement by the Managing Director of Lanerossi, an Italian Company which is the second largest wool textile manufacturer in the world. He commented that 'the over-riding constraint on the demand for wool was still the lack of stable prices' and added that 'price stability was even more important than prices as such' [4, p. 67].

Stabilisation schemes for wool exist in Australia, New Zealand and South Africa but they have not been effective in stabilising the price of wool although they may have reduced price fluctuations. The Australian scheme operated by the Australian Wool Corporation provides for a floor price for wool. The Corporation buys wool at auctions and stocks it if the auction price would otherwise fall below the floor price. The wool is then disposed of when market conditions improve. Thus a buffer stock system is in operation intended to place a floor to downward price movement but not intended to stabilise wool prices at some average level. Given the present marketing arrangements for wool it is still possible for its price to spiral upwards in periods of strong demand.

---

[3] Incidentally, the relative contributions of wool producing countries to Research and Development activity appear to support Olson's theory that in the provision of a collective good large gainers are liable to be exploited by the smaller ones. Australia's relative contribution appears to be greater than that of New Zealand which, in turn, seems to be greater than that of South Africa, and there are other wool producers who contribute very little or not at all to Research and Development for the wool industry. See Olson [26] and Olson and Zeckhauser [27].

Fluctuations or instability of wool prices have been claimed by the Australian Wool Commission to be extremely detrimental to the competitive position of wool. In its report on *The Marketing of Australian Wool*, The Australian Wool Corporation said:

> ... the most significant factor mitigating against wool's acceptance by the textile industry is the lack of assurance of its future availability at a predetermined price. Stated another way, users are unable to predict the prices they will pay to obtain the supplies of wool necessary to meet their forward commitments. Because of their direct method of selling, man-made fibre manufacturers have a distinctive competitive edge over wool in that they have the ability to ensure delivery of adequate supplies of their product to these industry leaders at a pre-determined price [4, p. 4].

In order to reduce uncertainty of wool prices, the Australian Wool Corporation proposed to the Australian Government in December 1973 that the Corporation should be the sole trader in all shorn wool for export. Since export wool amounts to 90 per cent of the Australian clip this would give the Commission a virtual monopoly position. Under the proposal, the Corporation would determine the quantities of wool offered for sale and the method of sale. "The Corporation would be able to manage the supply of virtually the entire Australian clip and determine the rate and method by which wool would be sold" [4, p. 6].

The Corporation envisaged that if this reform is adopted, (i) it will negotiate extended forward sales to key users and (ii) in times of high prices, when supplies are inadequate to dampen price rises, it will allocate or ration supplies to buyers at stable prices. The Wool Corporation report continued:

> The proposed system envisages forward contracts of up to 12 months to key users to ensure they are in a position to quote firm fabric and garment prices at a time when sample selections are being made. Sale by allocation is the proposed system's principal weapon against unacceptably high prices [4, p. 8].

During periods of relative stability of supply and demand for wool and thus relatively stable market prices for wool, the bulk of Australian wool might be expected still to be sold at auction. As at present, however, the Corporation would still set a floor price to transactions at the auction. The distinctive feature of the new proposals compared to the existing one is that the Australian Wool Corporation would become the sole trader in Australian export wool, would be prepared to enter into extended forward sales contracts for the supply of wool and would be prepared to set a ceiling price for wool and ration supplies if necessary.

Commenting in 1976 on the proposals of the Corporation, E.J. Donath, an agricultural economist, said:

> In a year or two a monopolistic Australian Wool Authority similar to the Australian Wheat Board will very likely be established with the power to acquire the whole Austr alian clip; this it will dispose of in either

> raw or processed form by any chosen combination of methods, including auction, tender, direct sale to the buying trade in Australia and direct sale to users. It is contended that only a monopolistic marketing organization can conduct a fully satisfactory supply management scheme integrating the many decisions necessary for wool-growers, processors and manufacturers [12, p. 99].

But, as Donath also points out, there is strong opposition from some wool brokers and some Australian topmakers to these proposals and there has been a swing in political opinion in Australia against market intervention by Government bodies. Consequently it is by no means certain that the Corporation's proposals will be implemented. The price of wool is likely to continue to fluctuate as political discussions about the proposals are delayed.

The Australian Wool Industry Conference, a body representing most Australian woolgrowers, supports the Australian Wool Corporation's 1973 plan for acquisition of the clip. The plan is also supported by the Wool Industry Policy Committee which is reportedly dominated by the Australian Wool and Meat Producers' Federation and Woolgrowers and Graziers' Council. The Government is, however, reported (early 1977) to be hesitant about adopting the plan for the Australian Wool Corporation's acquisition of all wool for export because among other things it wonders whether the Corporation has the experience and the ability to handle such a large purchase.

Consequently, in February 1977 the Corporation proposed to the Minister for Primary Industry, Mr Sinclair, that it be allowed to purchase 150,000 bales of wool direct from growers in the 1977-78 season.[4] This plan for direct purchase from growers, Limited Opportunity to Purchase Trial (L.O.P.T.) would enable the Corporation to build up experience in handling direct purchases and could enable it to prove that it can reduce the cost of disposing of the Australian clip.

The practice of by-passing auctions and buying directly from woolgrowers - "private treaty trading" - has grown in recent years. It was reported that in 1974-75 about 23 per cent of the Australian clip was sold outside the auction by private treaty. The National Council of Wool Selling Brokers has criticised private buying on the grounds that it fragments the industry and may add to handling costs but competition has forced brokers to commence private buying through subsidiary companies.

It is possible that brokers may oppose the Corporation's Limited Opportunity to Purchase Trial because it may reduce their private buying and auction business. Furthermore, if L.O.P.T. is successful, it could convince the Government that the Australian Wool Corporation can handle the whole Australian clip at a cost saving and demonstrate the ability of the Corporation to carry out its 1973 marketing plan. On the other hand if L.O.P.T. is a failure the 1973 plan is unlikely to be adopted.

---

[4] See Diegutis [11] and Tooth [40].

Political indecision continues and as reported in *The Australian* of 15 February 1977:

> Reformers fear loss of support from their own ranks. As prices recover - the greasy average for this season is forecast to rise to 190 cents a kilogram - so farmers' discontent at wasteful marketing methods wanes.
>
> A return to former apathy could get the Government off the farm-vote hook. Already Canberra has at least three other reform plans to play off the conflicting proponents.
>
> Brokers, buyers and Queensland graziers have each submitted detailed plans to set up new marketing structures. The longer the Canberra bureaucrats take to put these through cost/benefit studies, the greater the chance of wool price recovery and the less the grower urgency for Government action [40, p. 7].

Meanwhile, despite the existence of a (flexible) reserve price for Australian wool, wool prices fluctuate and remain uncertain.

The yearly average prices of wool, cotton and selected man-made fibres between 1963 and 1977 are indicated in Fig. 2.5. For wool and synthetic (non-cellulosic) fibres, prices fell during the 1960s. However, by 1972-73 the prices of all fibres were rising. These prices peaked in 1973-74 or in 1974 depending on the fibre and then fell due to the subsequent world recession and began to rise again as a consequence of rising costs of production. Note that the recovery of wool prices in the 1970s seems to have lead that of synthetic fibres.

While the pattern of prices in Fig. 2.5 is no doubt of general interest these prices need to be processed to provide (i) an accurate picture of the variability of wool prices compared to that of other fibres and (ii) information on the price of wool relative to that of other fibres.

It should be noted that Japanese f.o.b. export *values* have been used as an indication of the prices of man-made fibres. This is more desirable than using listed prices because manufacturers of synthetics sometimes allow (secret) discounts. On the other hand, domestic (non-export) prices may differ from export values. For example, they may be more stable than export prices. If this is so and since most of the production of synthetic fibres is not traded internationally, use of export values may exaggerate the apparent degree of variability of the prices of man-made fibres.

Profiles of the degree of variation in fibre prices from year to year between 1963-64 and 1976-77 are shown in Fig. 2.6. The percentage change in the average price of fibres in each year compared to the preceding year is graphed. Clearly, the relative variability of the prices of all fibres is greater in the 1970s than in the 1960s. *On average* wool prices show greater variability for the whole period than other fibre prices. The average between-year variation of wool prices is 27 per cent. This compares with 20 per cent for polyester, 18.1 per cent for cotton, 16.56 per cent for rayon, 15.15 per cent for acrylic and 12 per cent for nylon. As suggested earlier, Japanese export *values*

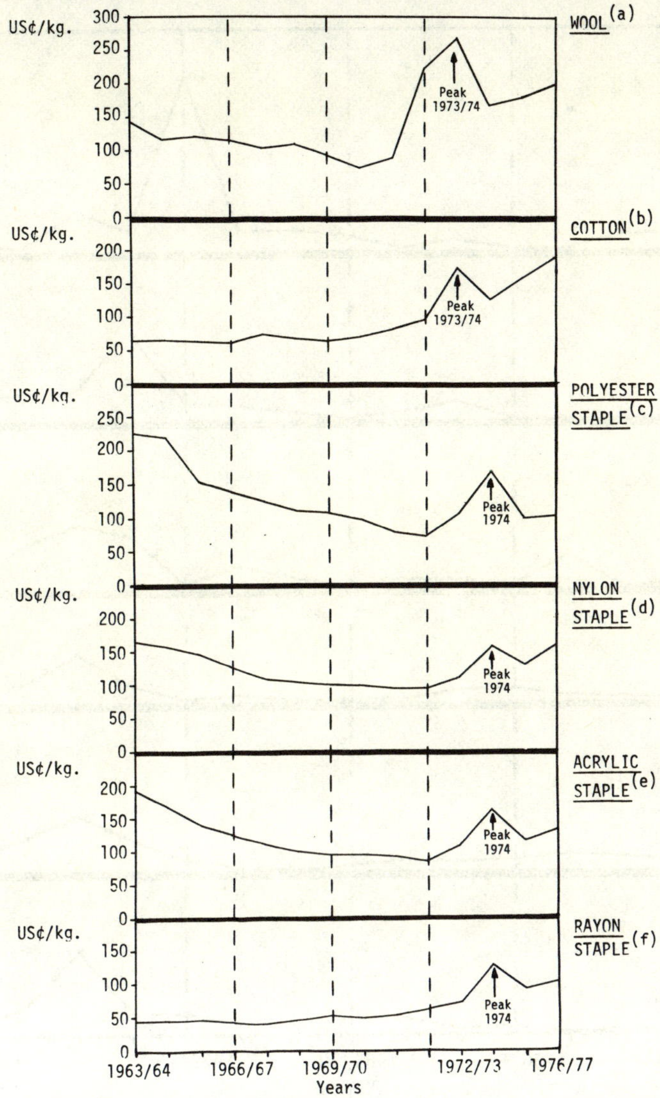

Fig. 2.5

Profiles of Average Yearly Fibre Prices 1963-64 to 1976-77 (US¢/kg.)

Notes: (a) Average greasy price, converted to US¢ at respective year end.

(b) Memphis Territory SM 1-1/16" C I F liverpool.

(c), (d), (e), (f) Japanese unit f.o.b. values. These prices are on a calendar year basis and stand for the first year shown.

Sources: Based on data supplied by The Australian Wool Corporation from data in *The Fibre Review* [6] (various issues) and from data in *Textile Organon* [33] (various issues).

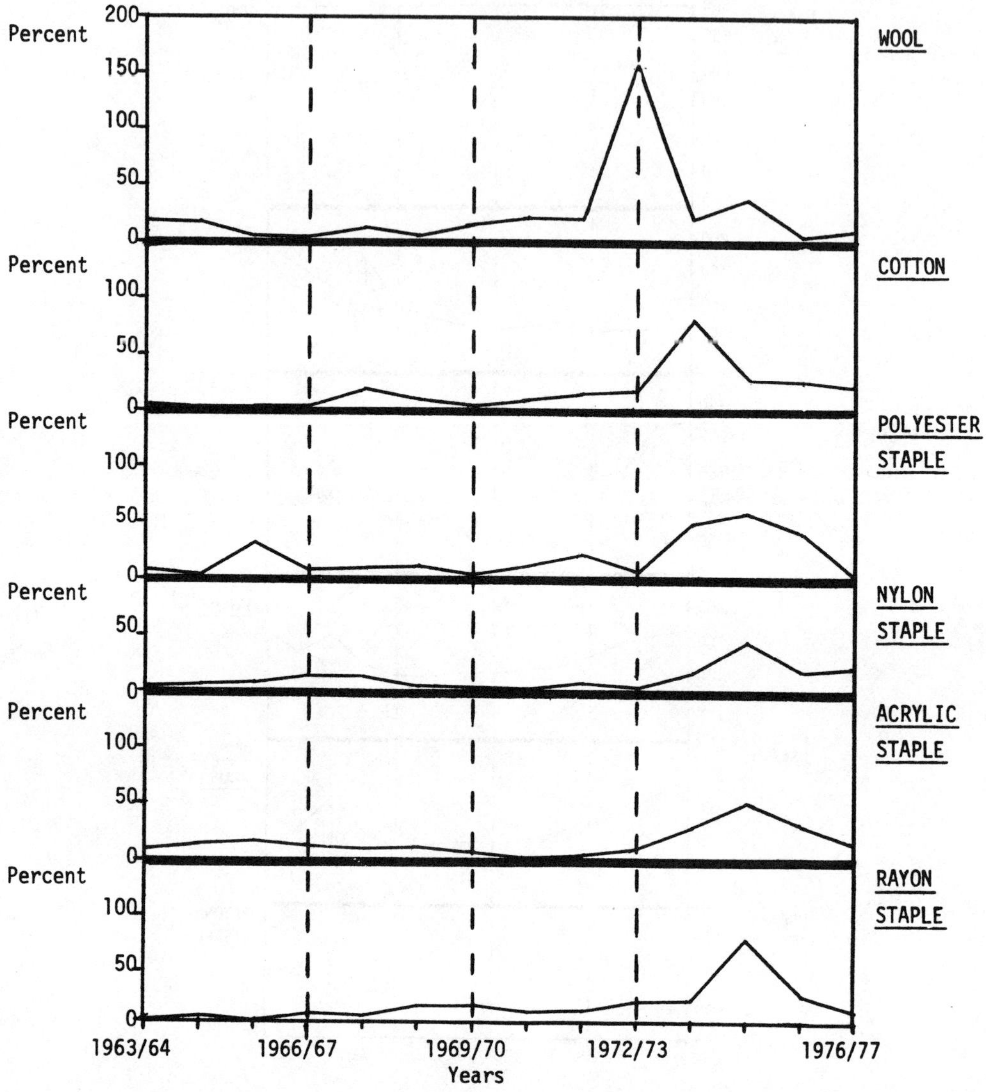

Fig. 2.6

Profiles of Between-Year Variability[(a)] of Fibre Prices[(b)] 1963-64 to 1976-77

Notes: (a) Expressed as a percentage on the vertical axis and measures the variation of the particular fibre price from its previous year's price.

(b) Same notes apply here as applied to Fig. 2.5.

Sources: Based on data supplied by the Australian Wool Corporation from data in *The Fibre Review* [6] (various issues) and from data in *Textile Organon* [33] (various issues).

for synthetic fibres probably exaggerate the degree of worldwide variability of the prices of man-made fibres. Domestic prices of man-made fibres are likely to be more stable and most of the volume of synthetic fibre production is not internationally traded.

It is clear that wool prices have shown the greatest relative instability but the relative fluctuations in the prices of all fibres have increased since 1971-72. During the textile boom of 1972-74 both cotton and wool prices rose relatively much higher and more quickly than the prices of man-made fibres. This phenomenon and associated factors will be given special consideration in the next three Chapters.

Another important factor affecting the market for wool is its price relative to other fibres, especially non-cellulosic fibres. The price of wool relative to that for the other main textile fibres is graphed in Fig. 2.7. The long-term trend is for wool prices to fall relative to cotton prices and rayon staple prices but to rise in price compared to the price of polyester, nylon and acrylic staple, the fibres most competitive with wool. The price of wool relative to the price of nylon almost doubled between 1962 and 1976; the price of wool relative to the price of acrylic more than doubled in the same period and it increased more than four times relative to polyester prices. Despite short-term perturbations, wool is becoming increasingly more highly priced relative to competing fibres. This clearly has implications for the promotion of wool.

The International Wool Secretariat predicts that the upward relative trend in wool prices in comparison to the price of synthetics will continue in the next few years, at least [37], [47]. Developments in the wool industry and in the synthetic fibre industry can be expected to bring this about. Over the next few years world wool production is expected to increase slightly (if at all) and this will tend to maintain wool prices. In the opinion of I.W.S. the prices of synthetic fibre will continue to remain depressed until after 1980. A recent paper by Wilcox, Area Director of I.W.S. in N.W. Europe, reported that:

> ... IWS could not see the efforts at self discipline by the [M.M.F.] industry succeeding in the near future against the political actions of some governments which would keep loss-leading producers alive to protect jobs. IWS could see continued investment in new capacity in Eastern Europe and the developing countries, and this would restrict the beneficial price lifting effect which might come from economic recovery in the West in say 1979 [37, p. 44].

But by the end of the 1980s rationalisation in the synthetic fibre industry and trade measures could reduce the over-capacity of the synthetic fibre industry.

> This should result, by 1987, in synthetic fibre prices reaching levels averaging between ten and 15 per cent higher in real terms than in 1977, with perhaps some manoeuvring of the wool price differential by then [37, p. 44].

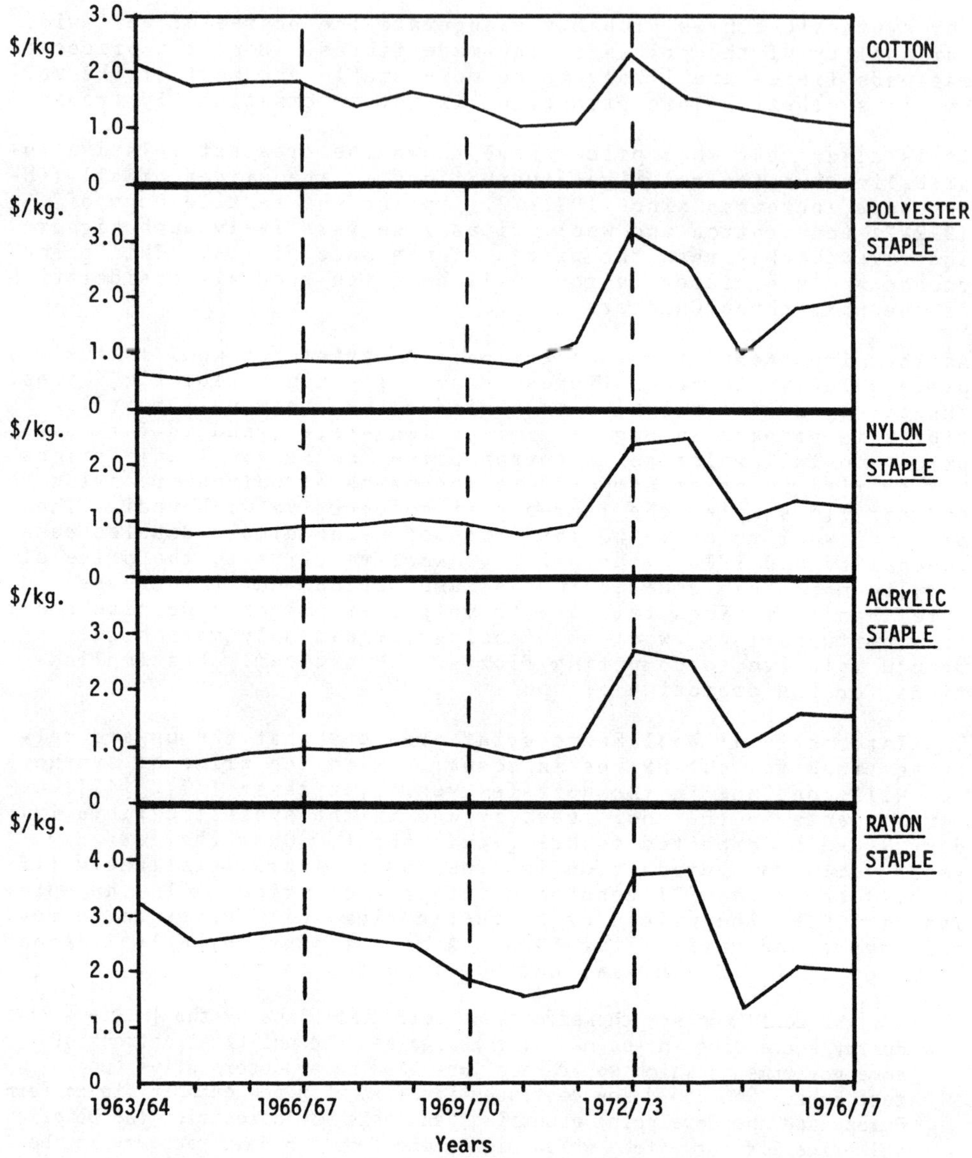

Fig. 2.7

Price of Wool Relative to the Price[(a)] of Other Main Textile Fibres 1963-64 to 1976-77 (per kilogram)

Note: (a) Same notes apply here for prices as applied to Fig. 2.5.

Sources: Based on data supplied by the Australian Wool Corporation from data in *The Fibre Review* [6] (various issues) and from data in *Textile Organon* [33] (various issues).

But rationalisation in the synthetic fibre industry may occur more quickly than anticipated by Wilcox. Europe's leading manufacturers of synthetic fibres are reported [25] to have agreed to a market-sharing scheme which will protect their relative market shares as they reduce capacity. It is expected that under the operation of the agreement capacity utilisation will rise to 85 per cent from the current level of 60-70 per cent. The main European synthetic fibre companies are participating in the scheme. These include I.C.I. and Courtaulds of Britain, Rhône-Poulenc of France, Montedison of Italy and Enka (textile subsidiary of Akzo), Hoechst and Bayer of Germany. American synthetic fibre companies in Europe while not formally parties to the reported cartel (otherwise they would be in violation of U.S. antitrust laws) are, it is said, "not expected to rock the boat".

While cartels are normally not legal under the rules for promoting competition within the E.E.C., there are special provisions in the Treaty of Rome which allow their formation under emergency conditions. Given the large losses incurred by European suppliers of synthetic fibres over the last few years and the serious over-capacity problem of the industry, special circumstances apply. To further consolidate the position of the European synthetic fibre industry, E.E.C. officials are expected to negotiate restrictions on the import of synthetic fibres to the E.E.C.

In Japan it is possible that legislation will be introduced to enable mergers to take place which will reduce the eight major synthetic fibre producers from eight to four [36]. It is anticipated that Asahi Chemical will merge with Kanebo, Teijin will join Unitika, Toray will link with Kuraray and Toyobo will merge with Mitsubishi Rayon. As a result of these mergers a 20 per cent reduction in productive capacity is anticipated.

This trend of a comparative rise in wool prices, unless arrested, seriously disadvantages wool. It has been claimed by the Economics Manager of the Australian Wool Corporation that relative fibre prices are of greater significance than previously in determining sales of wool [44, p. 27]. Retailers are becoming more price conscious in setting guidelines for their orders. "The situation is that a retailer sets a price limit and if, for example, a pure wool product fails to meet that limit the manufacturer may be obliged to shift to a lower priced blend" [44, p. 26]. There is also evidence that retail sales persons may not be giving as much advice to customers on the selection of products as previously because of self-service and in this case price rather than quality may have a bigger role in influencing choice. Consequently Ward points out that:

> Relative fibre prices are of greater significance than previously and, in view of known retailer price constraints, this significance is likely to increase. It applies not only to the comparative levels of fibre prices, but also particularly to the variability of prices - and it is at the manufacturing stages that the longest commitment is made to an established price between the presentation of sample ranges and the receipt of volume orders. The Corporation has frequently stressed the view that the long-term demand for wool is in jeopardy if further steps

are not taken to reduce wide fluctuations in wool prices (the reserve price scheme has only partially achieved this goal) [4, p. 27].

2.6 CONCLUSIONS

Data about the competitive strategies of man-made fibre producers is difficult to obtain. However, available evidence indicates that as is common in oligopolistic industries, product competition is more important than price competition and that promotion is a major factor in the successful marketing of man-made fibre. The major innovations in the industry in this Century has been the introduction of non-cellulosic fibres. Nylon was introduced just before World War II and polyester and acrylics in the 1950s and all have captured a considerable portion of the fibre market. Production of these fibres is concentrated in the hands of about 30 producers, most of which are multinational companies producing a range of different products. Several of the leading producers of man-made fibres are large chemical or petrochemical producers and have entered the industry since World War II, that is since non-cellulosic fibres sometimes called chemical fibres, based on oil derivatives, have been introduced. Their entry has enabled these large chemical producers to grow, diversify, and integrate vertical operations.

Producers of natural fibres have attempted to counteract competition from man-made fibres. The cotton industry, it seems, was rather slower than the wool industry in responding to the increased competition from the man-made fibre industry following the successful introduction of non-cellulosic fibres. The expenditure of the cotton industry on research and promotion before the mid-1960s was relatively low and the policy of the United States, the world's major exporter of cotton, of stockpiling cotton and maintaining world cotton prices in an effort to maintain farm incomes, gave man-made fibre producers a relative advantage during the 1950s and early 1960s. From the mid-1960s onwards however, the United States altered its policy to one of keeping United States prices of cotton roughly in line with international supply and demand for cotton and measures were introduced to increase promotion of and research into cotton and its uses.

In the case of wool, major exporting countries had already foreseen in 1937 the need to co-operate in promoting wool and formed the International Wool Secretariat for this purpose. The need for the wool industry to expand its promotional activities to counter increasing competition from man-made fibres became apparent in the 1950s, and in the early 1960s the operations of the International Wool Secretariat were expanded considerably and the industry adopted some of the competitive strategies which had been used by makers of man-made fibre producers. While these strategies were useful in maintaining the demand for wool, they had little or no impact on the instability of wool prices, a factor which it has been claimed places wool at a competitive disadvantage compared to man-made fibres Although wool boards in the major wool producing countries have attempted to increase the stability of wool prices, considerable instability remains. The Australian Wool Corporation has contended that the elimination of this instability is necessary if wool is to compete successfully with man-made fibres and has proposed a scheme to stabilise the price of the Australian wool clip.

REFERENCES

[1] AIROV, J., *The Location of the Synthetic Fibre Industry*, Technology Press of the Massachusettes Institute of Technology and John Wiley, Cambridge, 1959.

[2] ALBACH, H., *Als-ob-Konzept und Zeitlicher Vergleichsmarkt: Zur Berechnung Hypothetischer Wettbewerspreise auf Oligopolistischen Markten*, J.C. B. Mohr (Paul Sieback), Tubingen, 1976.

[3] ALBACH, H., "Collusion and Competitive Equilibrium in an Oligopolistic Market", *The German Economic Review*, 15, 83-97, 1977.

[4] Australian Wool Corporation, *The Marketing of Australian Wool*, Melbourne, 1973.

[5] BERRER, M., *Speech*, 24th Plenary Meeting of the International Cotton Advisory Committee Symposium on Inter-Fibre Competition, Washington, D.C., 1965.

[6] Bureau of Agricultural Economics, *The Fibre Review 1975-76*, Australian Government Publishing Service, Canberra, 1977.

[7] Bureau of Agricultural Economics, "Outlook for Cotton", *National Agricultural Outlook Conference, 1978: Papers Presented*, Australian Government Publishing Service, Canberra, 240-245, 1978.

[8] Business Press Ltd., "Cotton and Its Future", *Textile Asia*, May, 104-110, 1976.

[9] Business Press Ltd., "Prospect of Decline: The International Competitiveness of Japan's Textile Industry", *Textile Asia*, September, 128-140, 1976.

[10] Business Press Ltd., "Wool Promotion is a Long Term Business", *Wool Record*, 132, 16, 1977.

[11] DIEGUTIS, P., "Costs Cloud Wool's Future", *Sydney Morning Herald*, March 14, 1977.

[12] DONATH, E.J., "Ways to Sell Wool", *Textile Asia*, February, 97-99, 1976.

[13] HAGUE, D.C., *The Economics of Man-Made Fibres*, Duckworth, London, 1957.

[14] HAKIM, O.A., "The Effects of United States Cotton Policy on the World Market for Extra-Long Staple Cotton", Unpublished Ph.D. dissertation, Department of Economics, University of Arizona, 1972.

[15] HOWRY, K.A., "Historical Profiles of U.S. Fibre Producers", *Modern Textiles*, 56, 20-78, 1975.

[16] HUFBAUER, G.C., *Synthetic Materials and the Theory of International Trade*, Duckworth, London, 1966.

[17] Industries Assistance Commission, *Draft Report on Financing Promotion of Rural Products*, Canberra, 1976.

[18] International Wool Secretariat, *Promotion in Action*, London, 1975.

[19] Japan Chemical Fibres Association, *Man-Made Fibres of Japan*, Tokyo, 1973.

[20] LIANG, L.S., "Marketing Organisation Strategy of Man-Made Fibre Producers in the U.S.A.", Paper presented at the Textile Institute Annual Conference, Sydney, February, 12-15, 1975.

[21] McNEIRNEY, F., "Ingenious Inquisitive Persons: 50 Years of Progress in the Man-Made Fibres", *Modern Textiles*, 56, 10-19, 1975.

[22] MAIDEN, A.C.B., "The Outlook for Wool", *National Agricultural Outlook Conference, 1974: Documents and Discussion, Vol. 2.*, Australian Government Publishing Service, Canberra, C50-C51, 1974.

[23] MARKHAM, J., *Competition in the Rayon Industry*, Harvard University Press, Cambridge, 1952.

[24] National Advisory Committee on Food and Fibre, *Cotton and Other Fibre Problems and Policies in the United States*, Washington, D.C., July, 1967.

[25] Newsweek Inc., "Fibre Producers Organise Cartel", *Newsweek*, May 15, 44, 1978.

[26] OLSON, M., *The Logic of Collective Action: Public Goods and the Theory of Groups*, Harvard University Press, Cambridge, 1965.

[27] OLSON, M., and ZECKHAUSER, R., "An Economic Theory of Alliances", *The Review of Economics and Statistics*, 48, 1966.

[28] RAMSEY, C., "Man-Made Fibre Marketing Concepts Relating to Innovation", *Cotton and Allied Textile Industries*, 15, 57-62, 1974.

[29] Rayon Publishing Corporation, *1975 Man-Made Fibre Yearbook*, New York, 1975.

[30] Rayon Publishing Corporation, "Historical Profiles of U.S. Fibre Producers", *Modern Textiles*, 56, 20-78, 1975.

[31] ROBSON, R., *The Man-Made Fibres Industry*, Macmillan, London, 1958.

[32] SCHUMPETER, J., *Capitalism, Socialism and Democracy*, George Allen and Unwin, London, 4th ed., 1954.

[33] Textile Economics Bureau, *Textile Organon*, New York (various issues).

[34] Textile Business Press, *Textile Month*, February, 19, 1976.

[35] Textile Business Press, "European Fibres: Ten-Year Nightmare", *Textile Month*, February, 17, 1978.

[36] Textile Business Press, "French Fibre Group to Trim Their Capacity in New Policy Move", *Textile Month*, February, 23-24, 1978.

[37] Textile Business Press, "Natural Fibres: Conference Looks at Future Prospects", *Textile Month*, February, 43-56, 1978.

[38] The Oriental Economist, *Japan Economic Yearbook 1975-76*, Tokyo, 1976.

[39] Time Inc., *Fortune*, May, 210-220, 1975 and August, 156-167, 1975.

[40] TOOTH, K., "Wool War Looms Again", *The Australian*, February 15, 1977.

[41] United Nations, "Problems of Raw Materials and Development: Declaration and Programme of Action", Centre for Economic and Social Information of the United Nations, E21, August, 1974.

[42] United States Department of Agriculture, *Developments in the Japanese Textile Industry*, May, 1975.

[43] United States Department of Agriculture, *Cotton and Wool Situation*, June, 1975.

[44] WARD, L.E., "The Future of Wool as a Textile Fibre", *Farm Policy*, 17, 24-29, 1977.

[45] WARD, L.E., "Wool: Implications of Development in the Synthetic Fibre Industry", *National Agricultural Outlook Conference, 1978: Papers Presented*, Australian Government Publishing Service, Canberra, 109-130, 1978.

[46] WILCOX, J., "The Position of Wool in the Market Place Over the Next Ten Years", a paper presented at a conference, *The Future of Natural Fibres*, Shirley Institute, Manchester, November, 1977.

[47] WILCOX, J., "Why Wool?", (mimeo), International Wool Secretariat, London, December, 1977.

[48] ZWICK, H., "Man-Made Fibre Marketing at Hoescht", *Cotton and Allied Textile Industries*, 15, 50-56, 1974.

CHAPTER 3

Variations in Natural Fibre Prices: Theoretical Effects of Market Penetration by Man-Made Fibres

3.1 BACKGROUND

The share of the total world fibre market held by man-made fibres has increased significantly. In 1962, man-made fibres accounted (by weight) for twenty-six per cent of the consumption of major textiles fibres (synthetics, cotton and wool) and by 1974 increased their share to forty-three per cent.[1] However, in that period the share of non-cellulosics (synthetic fibre made from oil and coal by-products) almost quadrupled. In recent years, both the share of synthetic fibres in total fibre consumption and the variety of man-made fibres have increased.[2] The question which arises is whether or not these developments, when considered in conjunction with the pricing and supply characteristics of synthetic fibres, have influenced fluctuations in the price of natural fibres, especially wool.

This matter has been raised in the literature before. In 1971 Tier and Kidman [14] undertook a systematic spectral study of price movements in and between the wool, wool tops and worsted yarn spot and future markets. They conclude amongst other things that:

> Total variance [of wool prices] around the long-term trend has also decreased over time, even after allowing for the lower levels of prices.[3] The pattern of decline in prices as between qualities of wool and the reduction in fluctuations in prices are believed to have been due largely to the competition from synthetics. [14, p. 80]

In their opinion, the principal cause of the fall in the variance of wool prices has been the development in recent years of new textile machinery which permits manufacturers of textiles to switch more easily between fibres. Substitution by manufacturers rather than by final consumers is seen as being most important in the short-term. While this factor of substitutability is likely to have an influence on wool price fluctuations, there are other

---

[1] For further statistics relevant to the trends in the fibre market refer to the Bureau of Agricultural Economics bulletin [2].

[2] See Chapters 1 and 2 for an account of these developments.

[3] This downward trend has not been maintained however. For evidence of this see Chapter 4.

factors to consider such as the nature of the supply of man-made fibres. The implications of changing supply conditions for fibres because of the growing importance of synthetics is considered in the model developed in this Chapter.

A more theoretical study of wool price fluctuations has been undertaken by Duane [3]. He contends, for various reasons, that in the last two decades:

> ... the price elasticity of demand for wool has been increasing; i.e. if we think of the above model as describing the wool market, there has been an increase in the absolute value of β [inverse of the slope of a Marshallian demand curve]. The effect of this is to decrease the variability of wool prices ...[4]

However, Duane's conclusion needs to be interpreted cautiously. In terms of his model, he shows that a rising price elasticity of demand is accompanied by a diminishing residual variance of wool prices [3, p. 109]. Duane only shows in his model that the unexplained variance of wool prices (degree of uncertainty or ignorance) declines with increases in the elasticity of demand for wool. One, therefore, wonders whether he is justified in asserting generally that because the price elasticity of demand for wool has fallen, wool prices are more stable and that there is less need for organisations (such as the Australian Wool Corporation) to intervene in markets to stabilise prices. Indeed, in a model which we consider below the effect is the opposite to that suggested by Duane - greater price elasticity of demand for wool, other things unchanged, intensifies fluctuations in wool prices.

3.2 ANALYSIS OF CHANGES IN THE SUPPLY CURVE OF FIBRES AND IN EQUILIBRIUM PRICES DUE TO THE GROWTH OF SYNTHETICS

The pattern of variations in wool prices may alter either because supply conditions in the fibre industry change, or because there are changes in the demand for fibres. Let us consider some implications of possible variations which might have occurred in the supply conditions in the fibre industry and then deal with implications for wool price variations of alterations in the demand for wool.

The analysis begins with the proposition that the supply conditions for fibres have altered because of the growing importance of man-made fibres as a proportion of total fibre use, and then explores the possible effects of this on the variation of fibre prices using Marshallian comparative static analysis. While such analysis is limited because it makes no allowance for lags and dynamic factors, it gives some useful leads for further analysis.

---

[4] See Duane [3, p. 8]. Following on from here Duane develops an empirical model of the wool market to estimate the effects of increase in price elasticity which might have occurred during the period of analysis.

Assume for theoretical purposes that man-made fibres and natural fibres are perfect substitutes, even though they are not perfect substitutes in practice. Imagine that the long-run supply curve of man-made fibres is constant[5] like curve LRS_M shown in Fig. 3.1 and suppose that the production of natural fibres involves long-run increasing cost so that the long-run supply curve of natural fibres is like curve LRS_N in Fig. 3.1. The long-run supply curves of man-made fibres and natural fibres are, therefore, quite different.

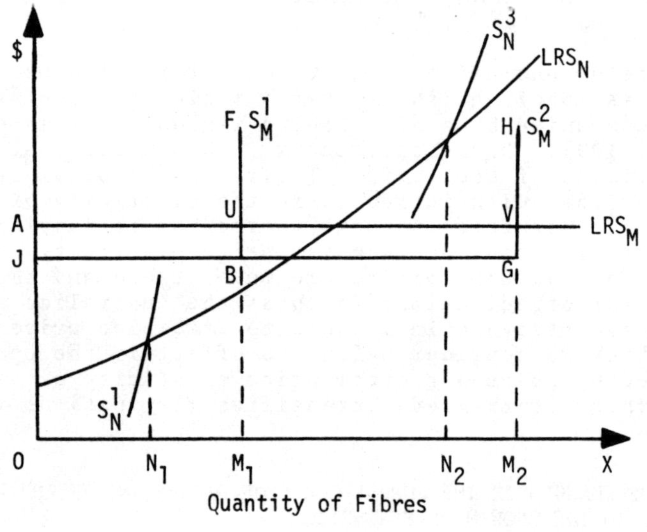

Fig. 3.1

Turning to the short-run supply curves for fibres, the extent to which these differ between natural and man-made fibres can be debated. However, since some research indicates that short-run supply curves in manufacturing industry tend to be of a reversed

[5] A constant long-run supply curve is not inconsistent with U-shaped long-run average cost curves nor with Bain-type long-run average cost curves which have a flat minimum range. Empirical evidence indicates that economies of scale occur in the production of synthetic fibres but these are not as marked as in some manufacturing industries. Pratten [12] using U.K. data estimates that at 50 per cent of optimum scale, per-unit costs of synthetic fibre manufacturers are five to seven per cent higher than at optimum scale. In respect of rayon plant economies, Markham [11] found that per-unit costs remain constant for plants in excess of a particular size.

L-shape,[6] assume that the supply curve of man-made fibres is of this shape. When the installed capacity of the man-made fibre industry is OM_1, in Fig. 3.1, the short-run supply curve of the industry may be like S_M^1, the right-angled curve JBF. When installed capacity is OM_2, the short-run supply curve may be like S_M^2, the curve JGH. In the natural fibre industry, by contrast, short-run supply curves are supposed to be normally sloped smooth curves but may be quite steeply sloped [4]. In Fig. 3.1, the short-run supply curve of natural fibres may be like S_N^1 when the normal supply of natural fibres is ON_1 and may be like S_N^3 when the normal supply of natural fibres is ON_2. Because of the greater number of farmers likely to be producing natural fibres when ON_2 is the normal production of natural fibres, S_N^3 may be less steeply sloped than S_N^1.

Using the information in Fig. 3.1, and the assumption that fibres are perfect substitutes, we can construct the long-run supply curve of all fibres. The long-run supply curve of combined fibres is KLT in Fig. 3.2, where the portion KL is the lower part of LRS_N in Fig. 3.1 and portion LT corresponds to LRS_M in Fig. 3.1.

Imagine that the fibre industry is in long-run equilibrium with X_1 being the normal output of natural fibres and $X_3 - X_1$ being the installed capacity of man-made fibre producers. The short-run supply curve of the fibre industry is then SS (dropping to the lower part of S_N^2 for very low prices) and is obtained by horizontally adding S_N^2 to the reversed-L supply curve of man-made fibres which corresponds to a man-made fibre capacity of $X_3 - X_1$. Thus the slope of the upper branch of SS is equivalent to the slope of S_N^2.

Suppose that the demand curve for fibres shifts (by equal vertical distances between curves D_1, D_2 and D_3 but has a central tendency to D_2. Given the market penetration of man-made fibres shown in Fig. 3.2, short-run equilibrium in the industry varies between *a*, *b* and *c*. However, in the absence of man-made fibres (since the dashed curve indicates the upper part of LRS_N and S_N^4

6 For evidence on the existence of a reverse-L marginal cost curve see Sherman [13, p. 255] and Johnston [10, p. 168]. Markham [11] specifically indicates that this type of short-run cost curve applies to the rayon producers. For the purposes of our theory, it is necessary only that the short-run supply curve be of an approximate reversed L-shape.

is the short-run supply curve of natural fibres corresponding to a supply of X_2) the short-run market equilibrium varies between *f*, *e* and *d*.

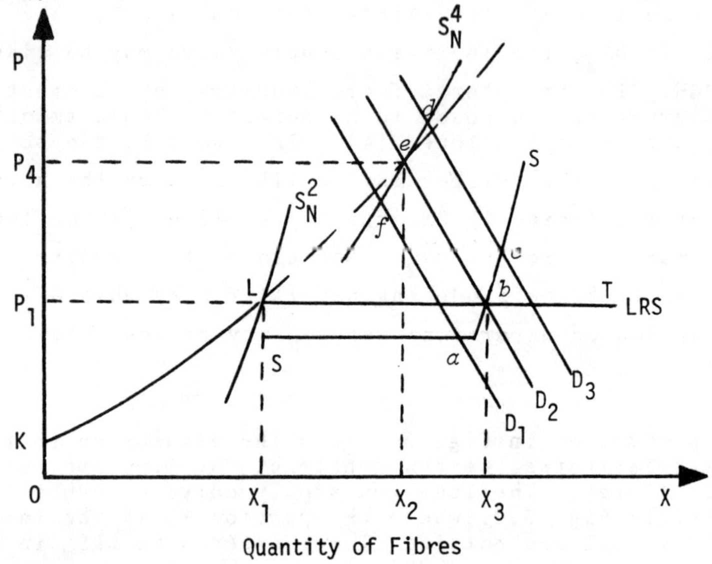

Fig. 3.2

The following conclusions can be drawn from this analysis. First, *relative upward* variations of price (relative to the appropriate measure of central tendency such as the mean or mode) are much greater once man-made fibres have penetrated the fibre market. Letting a fibre price of P_5 correspond to equilibrium *d* and P_2 correspond to equilibrium *c*

$$(P_2 - P_1)^2/P_1 > (P_5 - P_4)^2/P_4.$$

This is so for two reasons: (i) $(P_2 - P_1) > (P_5 - P_4)$ because S_N^4 is less steeply sloped than S_N^2 which has the same slope as the upper branch of SS), and (ii) $P_1 < P_4$. Relative variations of price above the mean fibre price may increase as man-made fibres penetrate the fibre market because the mean price of fibres falls and the upward stretch of the short-run supply curve of fibres may become more inelastic. Even if the latter effect is not important, the former one could be. Secondly, for downward variations of price from the mean price, these are lower *absolutely* when man-made fibres penetrate the market than before. If a fibre price of P_0 corresponds to *a* and P_3 to *f*, $(P_1 - P_0) < (P_4 - P_3)$. One cannot, however, conclude that downward variations of fibre prices will be lower relative to the mean (or modal) fibre price. They may be lower relatively but can be also higher depending upon the extent to which the mean price of fibres is lower once man-made fibres have penetrated the market.

In view of these considerations there is a strong possibility that the coefficient of variation of fibre prices tends to rise (other things equal) as market penetration by man-made fibres occurs.[7] Furthermore, if the long-run supply curve of man-made fibres tends to shift downward, for instance, as a result of technical progress, this can add to any tendency for an increase in coefficient of variation of fibre prices because the mean price of fibres is declining. Thus the growing importance of synthetic fibres in total fibre consumption may cushion natural fibres against very large declines in fibre prices but exaggerate any increase in prices when the demand for synthetics exceeds the productive capacity of manufacturers of man-made fibres.[8]

If the man-made fibres industry is oligopolistic [12] or even monopolistic and producers engage in full cost pricing,[9] the short-run supply curves of the industry do not fall below the long-run supply curve. Prices are inflexibly set to cover full costs including the normal return for the industry. Consequently, if, in Fig. 3.1 man-made fibre producers together have a total capacity of OM_1, the short-run supply curve of man-made fibres is AUF rather than JBF. Again, if the capacity of man-made fibre producers totals OM_2, the short-run supply curve of man-made fibres is AUF rather than JBF. Again, if the capacity of man-made fibre producers totals OM_2, the short-run supply curve of man-made fibres is AVH rather than JGH. Producers of man-made fibres hold their price constant at OA, until full capacity is achieved. Hence in Fig. 3.2, if the normal production of natural fibres is X_1 and man-made fibre capacity is $X_3 - X_1$, the short-run supply curve of all fibres combined is KL*bc*S. Under these conditions the earlier observations about relative upward variations of fibre prices away from the modal or central price still apply but downward variations are moderated.

We must now consider whether synthetic fibres rise in price once the demand for them exceeds the capacity of existing plants. The Australian Wool Corporation in a report [1] published in 1973 suggested that manufacturers of synthetic fibres tend to meet demand not by raising prices, but by rationing supplies if demand exceeds their productive capacity. Synthetic fibre prices seem to be very stable compared to natural fibre prices but to a considerable extent this stability in the past might have been a

---

7 This proposition is examined in more detail in Chapter 4.

8 An empirical investigation of the effects of man-made fibre capacity utilisation on the variability of natural fibre prices is undertaken in Chapters 5 and 6.

9 Full cost pricing has been given particular attention by Hague [6], [7] and [8].

reflection of excess capacity in the industry and constant (or even slightly falling) marginal costs of production in the industry. Natural factors, such as weather, also have little influence on the supply of synthetic fibres.

The Australian Wool Corporation [1] suggested that price stability plus possible rationing of supplies formed a part of the marketing strategy of manufacturers of synthetic fibres. However, evidence indicates that once full capacity is reached, as happened in 1973, the price of synthetic fibres does rise. Furthermore, with the expiry of patents and the widespread appearance of unbranded fibres, competition between manufacturers of synthetic fibres has increased. The suggestion that manufacturers of synthetics follow price stabilisation policies needs further investigation.

However, if manufacturers of synthetics do ration supplies once demand exceeds their productive capacity, the effect is to exaggerate the peak levels of natural-fibre prices even more than suggested in the above analysis. Manufacturers of synthetics by artificially holding down the price of synthetic fibres raise the profit and liquidity of textile manufacturers using synthetic fibres above the levels which these manufacturers would otherwise achieve under boom conditions. Because of this increased profitability and liquidity in periods of high demand, textile manufacturers could be expected during a boom to raise their demand for natural fibres to levels greater than otherwise would be the case. The time series of prices for natural fibres changes in the same way as suggested in the above analysis for all fibres but the price of natural fibres reaches a higher peak if the demand for synthetics at prevailing prices exceeds productive capacity. In the case of wool, one might expect the range of prices (or at least the upward range relative to the mean) to increase as the production of synthetic fibres becomes relatively more important.

3.3 EFFECT OF INCREASED ELASTICITY OF DEMAND ON PRICE FLUCTUATIONS

Having examined some important aspects of the supply of fibres, the demand for fibres, especially wool, is now considered. Many of the analytical points which are discussed for wool apply to cotton and hold generally.

In the literature [3], [14], it has been contended that the growing importance of synthetic fibres results in a more elastic demand curve for wool and this implies, other things being equal, a reduction in the fluctuation of wool prices. Increased elasticity has been interpreted by Duane [3], to mean that the demand curve for wool has less downward slope.

For the purpose of this section, let us *adopt Duane's interpretation of treating slope and elasticity as synonymous*. As is well known, elasticity and slope are not synonymous in reality. No confusion will, however, occur in the present analysis if it is remembered that slope is the *desideratum*.

It will be shown that under certain conditions, a reduction in the downward slope of the demand curves for a commodity increases price instability in the market for the commodity. This consequence, which is the opposite to that suggested by Duane, occurs when (i) the market demand curves maintain the same vertical difference between their height as their slope diminishes, (ii) market equilibrium is established as demand shifts, and (iii) the supply curve is neither perfectly elastic nor perfectly inelastic.

The result is illustrated in Fig. 3.3. Imagine that the demand curve is originally D_1 half of the time and D_2 the other half of the time. With the supply curve of commodity X (of wool, say) shown as S, equilibrium fluctuates between *b* and *d* and the average price of X corresponds to point *c*. Now let the slope of the demand curves fall in a manner which keeps their difference in height, HJ constant and maintains the average equilibrium price of X unaltered. This is achieved by drawing the new demand curves parallel to one another through points H and J.[10] The most elastic possibility is shown in Fig. 3.3 by the demand curves D_1' and D_2'.

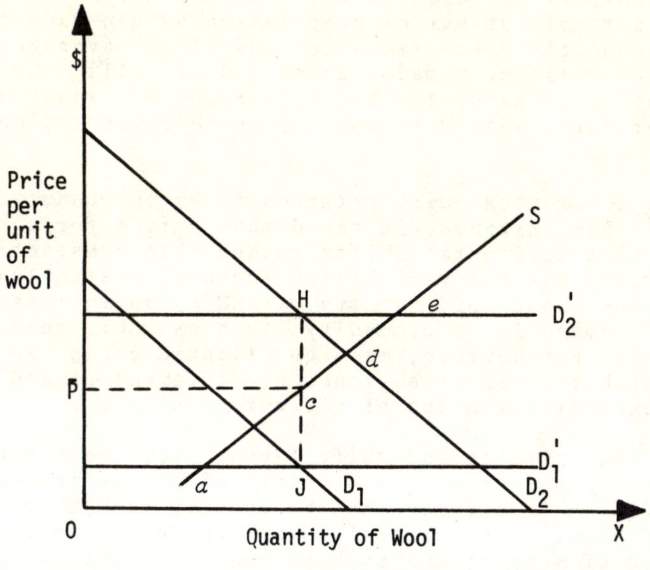

Fig. 3.3

Given these elastic demand curves which exhibit the same shift factor as in the previous situation, equilibrium fluctuates between point *a* and point *e* and price fluctuations increase accordingly. The average market situation remains unaltered at *c*. It

---

[10] H and J are points such that *a*J = *e*H. All demand curves which leave the average price of the product unchanged at *c* and show a shift equivalent to distance HJ must pass through points H and J irrespective of their slopes.

is readily seen, and can be shown rigorously[11] that as the absolute slope of the demand curve becomes smaller (as it becomes more elastic), other factors unaltered, price and quantity fluctuations in the market for commodity X (wool) increase.

In the case just analysed, it was assumed, as a part of holding conditions constant, that after the change in the elasticity of demand, average price is unaltered. If this is so, then under the conditions postulated, average profits from wool rise, as a result of an increase in the elasticity of demand for it [15], [16].

Indeed, under the above conditions, increased price fluctuations arise from reducing the absolute slope of the demand curve irrespective of whether the same average price for the commodity is maintained. This result holds if the vertical shift of the demand curve remains constant and the supply curve is neither perfectly elastic nor perfectly inelastic.

The instability of wool prices (or of any other commodity) rises in the above circumstances if the elasticity of its demand curve increases. Thus, if the elasticity of demand for natural fibre increases as a result of market penetration by man-made fibres, if the above conditions are satisfied and if the average price of natural fibres either remains unchanged or falls, the coefficient of variation of natural fibre prices can be expected to rise. There is, therefore, a need to qualify Duane's theoretical conjecture.

But there are theoretical circumstances in which Duane's conjecture holds.[12] For instance, if the demand curves for a commodity maintain constant horizontal shifts rather than constant vertical shifts (thus also holding variability factors constant) and the other conditions above hold *mutatis mutandis*, an increase in the elasticity of demand for a commodity, such as wool, reduces its price variance. Furthermore, more complicated cases can be envisaged in which the relative slopes of different demand curves in the frequency distribution of the curves alter.

In the above analysis, demand-shift factors were held constant. But in practice the degree of and nature of shifts in the demand for natural fibres is affected by the growing market dominance of man-made fibres. The partial demand curve for wool depends upon the price of substitutes such as man-made fibres. If the elastic stretch of the short-run supply curve of man-made fibres (see Fig. 3.1) keeps the price of fibres from rising and from falling as the demand for fibres alters, the demand curve for wool will neither rise as high or fall as low as would be the

---

[11] This follows because the tan of angle *e*HD and of angle *a*J*e* must decline as the absolute slope of D_1 and D_2 falls and reduces the height of the triangles *e*Hd and *a*Jb. As the height of these triangles falls, the variation of equilibrium price about the average price corresponding to *c* increases.

[12] Pointed out by T. Tier in private correspondence.

case if the whole supply of fibres comes from natural sources. This effect (the elastic stretch of the supply curve of man-made fibres) will tend to moderate any increase in the variance of wool prices which might flow from a rise in the elasticity of demand for wool. However, the demand for fibres sometimes may be such that the demand for synthetic fibres cannot be met from productive capacity and equilibrium is established outside the elastic stretch of the supply curve of man-made fibres. On these occasions, the demand curve for wool may be higher than otherwise, could be quite inelastic and the price of wool could move to higher levels. Overall the coefficient of variation of wool prices may rise.

There are also other factors which may alter variations in the demand for wool and fluctuations of fibre prices. Man-made fibre production is less subject to uncontrolled disturbances, such as weather variations, than is natural fibre production. Total fibre supply might be expected to be less subject to random disturbance of a seasonal type as the market penetration of man-made fibres increases. In addition, firms which produce man-made fibres are generally larger than those which produce natural fibres. Generally, it is more profitable for larger firms to collect market information than for small firms to do so and, therefore, producers of man-made fibres may be better informed about market prospects than wool growers and cotton growers. Better information on the part of manufacturers of man-made fibres and a greater control of production may mean that fibre prices become less random as the market share of man-made fibres increase. This may be a factor operating in favour of price stability of man-made fibres.

On the other hand, other factors may contribute to greater instability in fibre markets. The consumption of fibres has been rising and much of this rise has been for income-elastic commodities such as carpets. Fluctuations in the general level of economic activity and incomes might be expected to generate significant swings in the demand for such commodities and add to swings in the cycle of demand for fibres. The importance of this change and its implications for wool require further analysis but with fluctuations of general activity rising, it could add to cyclical instability of demand for wool. A recent F.A.O. report [4] indicates that this income-elastic factor has an important effect on swings in the consumption of such synthetics as nylon, which are in heavy demand for industrial purposes and for use in consumer durables such as carpets.[13, 14]

---

[13] See [4] for further information on income and demand.

[14] Work by Hussey [9] indicates that over the nine year period ended first quarter 1972, most of the variation in Australian wool prices was associated with variations in interest rates in the main countries consuming Australian wool. He suggests that higher interest rates reflect increasingly restrictive government policies, falling economic activity, and therefore a decline in the demand for wool. This suggests that increased variability in management of aggregate demand by governments can raise the instability of wool prices by destabilising the demand for wool.

The net effect of all the above factors as far as wool prices and the price of natural fibres are concerned is uncertain.[15] But it is theoretically plausible that the coefficient of variation of the prices of natural fibres has risen as a result of market penetration by man-made fibres. Furthermore, natural fibres may become more variable in price as capacity utilisation in the man-made fibre industry increases. Trends in the variation of natural fibre prices will be explained in the next Chapter and the hypotheses just mentioned will be tested in subsequent Chapters.

REFERENCES

[1] Australian Wool Corporation, *The Marketing of Australian Wool*, Melbourne, 1973.

[2] Bureau of Agricultural Economics, *The Fibre Review 1973-74*, Australian Government Publishing Service, Canberra, 1975.

[3] DUANE, P., *Analysis of Wool Price Fluctuations*, Wool Economic Research Report No. 23, Bureau of Agricultural Economics, Canberra, 1973.

[4] F.A.O. Commodities and Trades Division, *Impact of the Energy Crisis on the Competitive Position of Agricultural Raw Materials*, Rome, 1974.

[5] GRUEN, F.H., and others, "Changes in Supply of Agricultural Products", in D.B. Williams (ed.), *Agriculture in the Australian Economy*, Sydney University Press, Sydney, 1967.

[6] HAGUE, D.C., "Economic Theory and Business Behaviour", *Review of Economic Studies*, 16, 144-57, 1949.

[7] HAGUE, D.C., *The Economics of Man-Made Fibres*, Duckworth, London, 1957.

[8] HAGUE, D.C., *Pricing in Business*, George Allen and Unwin, London, 1971.

[9] HUSSEY, D.D., "An Alternative Approach to Projecting Short-Term Wool Prices", *Quarterly Review of Agricultural Economics*, 25, 175-181, 1972.

[10] JOHNSTON, J., *Statistical Cost Analysis*, McGraw-Hill, New York, 1960.

[11] MARKHAM, J.W., *Competition in the Rayon Industry*, Harvard University Press, Cambridge, 1952.

[12] PRATTEN, C.F., *Economies of Scale in Manufacturing Industry*, Cambridge University Press, Cambridge, 1971.

[13] SHERMAN, R., *The Economics of Industry*, Little, Brown and Company, Boston, 1974.

[14] TIER, T.J., and KIDMAN, P.R., "Price Movements In and Between the Wool, Wool Tops and Worsted Yarn Spot and Futures Markets", *Quarterly Review of Agricultural Economics*, 24, 63-81, 1971.

---

[15] Some of the effects of the 1973 oil crisis are considered in [4]. It should be noted, however, that Japanese speculation did contribute in part to the significant fluctuations in wool prices that occurred in late March and early April 1973. Market reports indicated that a price decline of over forty per cent during this period was associated to some extent with a sharp reduction in Japanese wool purchases.

[15] TISDELL, C.A., *The Theory of Price Uncertainty, Production and Profit*, Princeton University Press, Princeton, 1968.

[16] TISDELL, C.A., "Some Circumstances in Which Price Stabilisation by the Wool Commission Reduces Incomes", *The Australian Journal of Agricultural Economics*, 16, 94-101, 1972.

CHAPTER 4

Past Trends in the Variability of Wool and Cotton Prices

4.1 INTRODUCTION

In order to decipher the main patterns of instability in wool and cotton prices, fibre prices for the 23-year period 1952-1974 were collected and their coefficients of variation were calculated. The purpose was not to give a definitive test of the hypotheses considered in Chapter 3, but to discover whether or not they could be rejected reasonably on the basis of even casual observations.

Our statistical observations cover a longer period than that considered by Duane [2] and Tier and Kidman [4] who examined trends in the fluctuations of wool prices. In the approach taken below very short-term fluctuations of wool and cotton prices (variations of monthly prices within years) and short-term variations (fluctuations of annual prices within five-year periods) are examined. In general, the statistical results are not inconsistent with the hypothesis (based on changed supply relationships) which was discussed.

The calculations for wool are based upon the average monthly prices of 21μ clean combing wool at Australian auctions. The price per kilogram for this wool is in Australian cents. The calculations for cotton are based upon Memphis Territory SM 1-1/16" CIF Liverpool prices expressed in U.S. cents per kilogram. The difference in the currencies used to indicate wool and cotton prices is inconsequential for our calculations because the coefficient of variation of prices is a relative measure of variability. The trends in prices for both wool (since 1949) and cotton (since 1952) are graphed in Fig. 4.1.

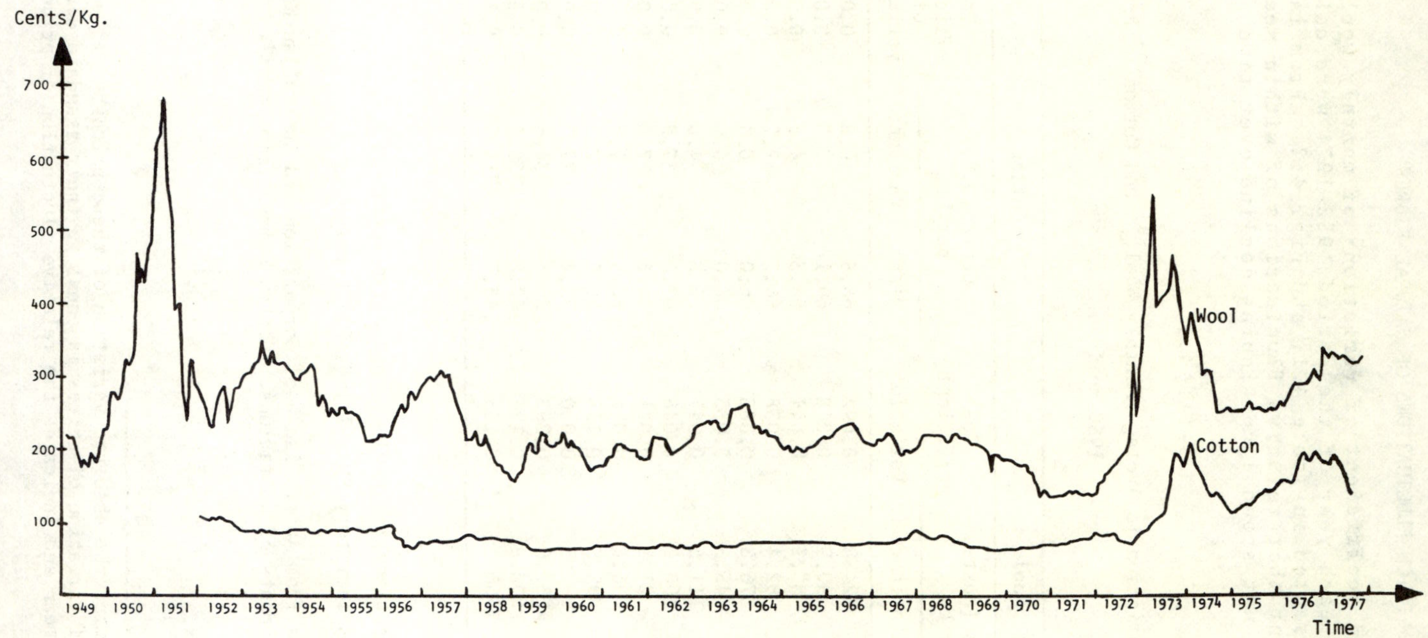

Fig. 4.1

Monthly Time Series of Wool[(a)] and Cotton[(b)] Prices 1949 to 1977

Notes: (a) 21μ clean combing wool at Australian Auctions.

(b) Memphis Territory SM 1-1/16" CIF Liverpool. Cotton prices begin 1952 - U.S. cents per kg.

Source: Wool Prices based on Australian Wool Corporation and Australian Wool Board Statements of quotations. Cotton Prices from *Textile Organon* [3].

4.2 VERY SHORT-TERM PRICE FLUCTUATIONS OF NATURAL FIBRES

The variance[1] and coefficient of variation[2] of monthly wool and cotton prices in each year of the period 1952-1974 were calculated and are presented in Table 4.1, and Fig. 4.2. From inspection it is evident that relative fluctuations of within season wool prices have not shown a continuing decline over time.

TABLE 4.1

Within-Year Variation in Average Monthly Wool[(a)] and Cotton[(b)] Prices 1960 to 1974

Year	Wool			Cotton		
	Cents (Aust.) per kg		Co-efficient of Variation	Cents (U.S.) per kg		Co-efficient of Variation
	Mean	Variance		Mean	Variance	
1960	196.9	324.8	0.092	66.5	0.8	0.013
1961	197.4	90.9	0.048	69.1	2.3	0.022
1962	206.3	83.3	0.044	68.3	4.6	0.031
1963	236.6	82.4	0.038	66.3	9.7	0.047
1964	233.6	376.8	0.083	66.0	0.0	0.000
1965	204.9	45.5	0.033	66.0	0.0	0.000
1966	223.8	111.6	0.047	63.8	4.0	0.031
1967	207.8	100.0	0.048	67.6	37.0	0.090
1968	216.8	21.1	0.021	74.3	17.8	0.057
1969	200.6	176.8	0.066	63.5	1.5	0.019
1970	168.6	395.5	0.118	66.2	2.5	0.024
1971	138.9	7.5	0.020	75.4	27.7	0.070
1972	216.8	4038.6	0.293	80.1	34.6	0.073
1973	419.9	2808.8	0.126	142.8	1839.2	0.300
1974	292.1	1566.1	0.135	143.7	714.1	0.185

Notes: (a) 21µ clean combing wool at Australian auctions.

(b) Memphis Territory SM 1-1/16", CIF Liverpool.

Sources: Wool prices from Australian Wool Corporation and Australian Wool Board statements of quotations. Cotton prices from *Textile Organon*, [3].

1 The variance measures the absolute variation of these prices.

2 The coefficient of variation of prices is equal to their standard deviation divided by their mean and indicates the relative fluctuation of prices.

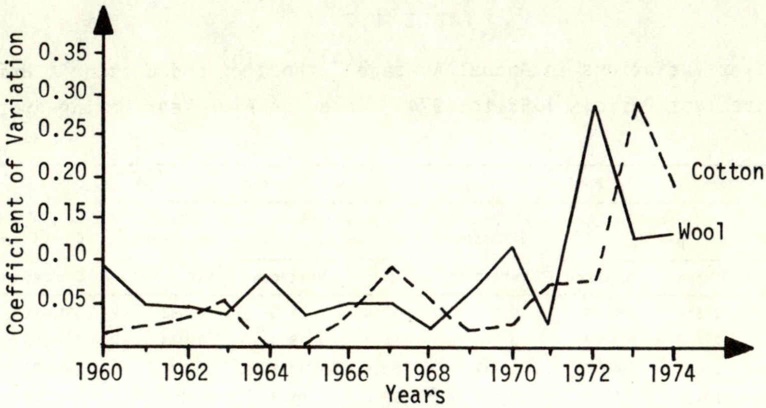

Fig. 4.2

Within-Year Coefficient of Variation of Wool and Cotton Prices 1960 to 1974

Source: Based on Table 4.1.

On the whole there seems to have been some tendency for a decline to occur until towards the end of 1968 and since then the coefficient of variation has risen. Thus observations over a longer period seem to call for a moderation or alteration of Tier and Kidman's hypothesis [4] that seasonal fluctuations in wool prices are becoming smaller. Certainly, this matter requires further and more sophisticated analysis. Calculations also indicate that the trend in the relative fluctuations of cotton prices is broadly the same as for wool.

The results suggest the hypothesis that the extent of seasonal variation in the prices of natural fibres may be linked with the presence of excess capacity in man-made fibre production.[3] There was considerable excess capacity in the 1960s for man-made fibre production and during this period relative seasonal variation in wool and cotton prices is lower than for other periods.

4.3 SHORT-TERM PRICE FLUCTUATIONS OF NATURAL FIBRES

Fluctuations in the short-term are relevant to testing the hypothesis introduced in Chapter 3. Consequently, variations in annual average wool and cotton prices based on five-year moving averages for the period 1952-1974 are presented in Table 4.2.

---

[3] The results shown in Table 4.1 when updated for 1975 and 1976 give the following coefficients of variation indicated in brackets:
Wool: 1975 (0.096), 1976 (0.097)
Cotton: 1975 (0.091), 1976 (0.098).
These are lower coefficients of variation than in 1974. This would be anticipated by our theory in view of greater excess capacity in the man-made fibre industry in 1975 and 1976 (see footnote 3, Chapter 5).

TABLE 4.2

Between-Year Variations in Annual Average[(a)] Wool[(b)] and Cotton[(c)] Prices During Five Year Periods 1952 to 1974 - Based on Five Year Moving Average

Years	Wool			Cotton			Coefficient of Correlation
	Cents (Aust.) per kg		Co-efficient of Variation	Cents (U.S.) per kg		Co-efficient of Variation	
	Mean	Variance		Mean	Variance		
1952-1956	272.6	1021.3	0.117	89.4	58.3	0.085	0.051
1953-1957	274.8	1014.7	0.116	84.4	30.3	0.073	0.003
1954-1958	250.0	1349.5	0.147	82.6	42.8	0.079	0.177
1955-1959	231.8	1276.7	0.154	78.6	68.8	0.106	0.316
1956-1960	223.6	1485.8	0.172	78.8	35.2	0.080	0.552
1957-1961	212.8	1329.2	0.171	71.6	25.3	0.070	0.468
1958-1962	198.4	18.8	0.022	70.0	20.5	0.065	0.433
1959-1963	206.6	285.3	0.082	67.6	1.3	0.017	0.763
1960-1964	214.0	381.5	0.091	67.2	1.7	0.019	0.825
1961-1965	215.6	326.3	0.084	67.0	2.0	0.021	0.734
1962-1966	221.0	221.0	0.067	66.0	2.0	0.021	0.428
1963-1967	221.4	206.8	0.065	66.0	2.0	0.021	0.393
1964-1968	217.6	140.3	0.054	67.6	14.8	0.057	0.202
1965-1969	211.0	87.5	0.044	67.2	17.2	0.062	0.232
1966-1970	203.8	454.7	0.105	67.2	12.2	0.062	0.224
1967-1971	186.8	1041.2	0.173	69.4	23.8	0.070	0.277
1968-1972	188.6	1152.8	0.180	71.8	44.2	0.093	0.170
1969-1973	229.2	12277.2	0.483	85.8	1072.3	0.383	0.958
1970-1974	267.3	12668.1	0.455	101.6	1467.9	0.376	0.836

Notes: (a) Simple average of monthly prices per kg in cents.

(b) 21μ clean combing wool at Australian auctions.

(c) Memphis Territory SM 1-1/16" CIF Liverpool

Sources: Wool prices based on Australian Wool Corporation and Australian Wool Board statements of quotations. Cotton prices from *Textile Organon* [3].

The coefficient of variation of annual average wool prices is graphed in Fig. 4.3. The coefficient rose until the five-year period ended 1961 and broadly fell during the 1960s until 1969. Since that time it has risen dramatically and appears to be more erratic. These results accord well with a hypotheses of Chapter 3 that when man-made fibre production approaches full capacity, variability of wool prices may rise. Excess capacity in man-made fibre plants was a feature of the early 1950s and the period 1969 to 1974.[4] As can be seen from Fig. 4.3, the trend in the variation of cotton prices has been broadly similar to that for wool. However, rising fluctuations in cotton prices appear to have led those for wool in the 1960s. The reason for this is not clear.

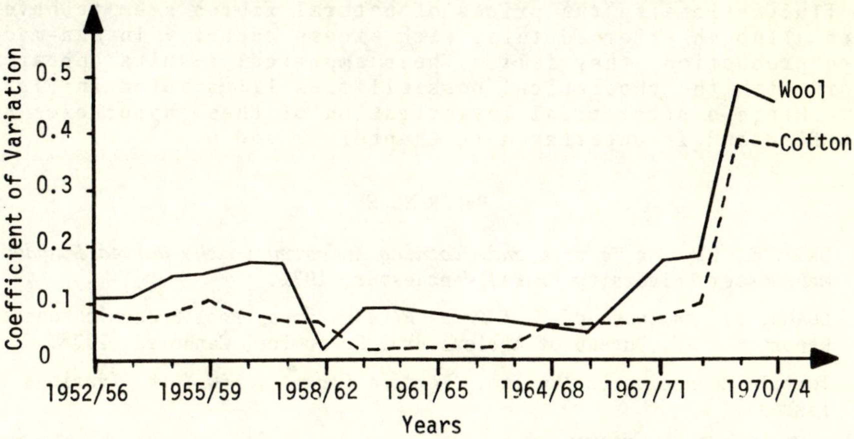

Fig. 4.3

Between-Year Variations in Annual Average Wool and Cotton Prices During Five Year Periods 1952 to 1974 - Based on Five Year Moving Averages[(a)]

Note: (a) Time in five year overlapping intervals.

Source: Based on Table 4.2.

It is contended that the historical evidence is sufficiently strong to reject the hypothesis of Tier and Kidman [4], that fluctuations in wool prices are declining - in recent years fluctuations have increased.[5, 6] The observed results are not inconsistent with the theory advanced in Chapter 3.

4.4 CONCLUSION

Historical results indicate that some accepted conclusions about wool price fluctuations call for reconsideration. On the face

---

[4] See Briscoe [1, Ch. 3].

[5] As mentioned earlier (Chapter 3, footnote 15) Japanese speculation did contribute in part to the significant fluctuations in wool prices in early 1973.

[6] The results shown in Table 4.2 when updated for 1975 and 1976 give the following coefficients of variation indicated in brackets:
Wool : 1971-75 (0.392), 1972-76 (0.818)
Cotton : 1971-75 (0.259), 1972-76 (0.492).
These are lower coefficients of variation than in 1970-74. This would be anticipated also by our theory in view of greater excess capacity in the man-made fibre industry in 1975 and 1976 (see footnote 3, Chapter 5).

of it observations seem to support the hypothesis of Chapter 3. Despite the rising share of fibre market accounted for by man-made fibres, relative fluctuations in wool and cotton prices have tended to increase. Furthermore, as man-made fibre production has approached full capacity (from the late 1960s onward) relative fluctuations in the prices of natural fibres seem to have risen although prior to this, with excess capacity in man-made fibre production, they fell. These empirical results appear to accord with the theoretical possibilities illustrated in Fig. 3.2. Hence a statistical investigation of these hypotheses seems justified and is undertaken in Chapters 5 and 6.

REFERENCES

[1] BRISCOE, L., *The Textile and Clothing Industry in the United Kingdom*, Manchester University Press, Manchester, 1971.

[2] DUANE, P., *Analysis of Wool Price Fluctuations*, Wool Economic Research Report No. 23, Bureau of Agricultural Economics, Canberra, 1973.

[3] Textile Economics Bureau Inc., *Textile Organon*, New York, (various June issues).

[4] TIER, T.J., and KIDMAN, P.R., "Price Movements In and Between the Wool, Wool Tops and Worsted Yarn Spot and Futures Market", *Quarterly Review of Agricultural Economics*, 25, 63-81, 1974.

CHAPTER 5

Instability of Wool and Cotton Prices Related to Capacity Utilisation of Man-Made Fibres: Single Regression Analysis

5.1 INTRODUCTION

In Chapter 3 we considered the theoretical possibility that the relative variability of wool and cotton prices is affected by the growing dominance of man-made fibres in the fibre market and concluded that swings in the relative prices of natural fibres may be intensified by the growing market share of man-made fibres. Basically, the earlier discussion relied on the proposition that the supply curve of fibres (or at least its relevant part) becomes more (reversed) L-shaped in nature as the market share of man-made fibres rises.

It was also speculated that the relative variability of natural fibre prices rises with increasing capacity utilisation in the man-made fibre industry. If the fibre supply curve is of a reversed-L character (but not necessarily strictly of an L-form - say of an exponential nature) and if natural and man-made fibres show a high degree of substitution at the margin of use,[1] alterations of a demand for fibres should result in little variation of natural fibre prices *when considerable excess capacity exists in the man-made fibre industry*. In this case, the market is attempting to reach equilibrium on the lower part of the L-shaped curve. However, as full capacity is approached in the man-made fibre industry equal swings in the demand curve for fibres result in considerable variations of natural fibre prices as the rising branch of the fibre supply curve is encountered.

Although past trends outlined in the last Chapter seem to support these hypotheses, specific testing of them is required. The relationships which we feel may be most marked, namely, a tendency for the relative variability of wool and cotton prices to rise with increased capacity utilisation in the man-made fibre industry, is explored in this Chapter using single regression analysis.

---

[1] It has been shown by Minford [3] that high degrees of substitution do exist between natural and man-made fibres in various end-uses.

In our empirical analysis, data estimates of the Textile Economics Bureau, published in *Textile Organon* [5], for annual world production and production capacity are used by us to calculate capacity utilisation in the man-made fibre industry.[2] To test our hypothesis that the relative variability of wool and cotton prices rises with increasing capacity utilisation in the man-made fibre industry, relative variations of these wool and cotton prices are related (using single regression analysis) to annual capacity utilisation figures for 1960-1974. The broad pattern of world capacity utilisation in the man-made fibre industry for the period 1960-1974 is shown in Fig. 5.1.

During the period 1960-1974 non-cellulosic capacity has been less fully utilised than cellulosic capacity. This may reflect the stagnation and relative decline of cellulosic production and rapid growth of non-cellulosics for which new plants have been built with excess capacity to tap expanding demands [2]. Throughout the 15-year period, capacity utilisation has varied. It was high in 1962 and 1973 and low in 1967 and 1974.[3] At no time was 100 per cent aggregate utilisation achieved.

However, bottlenecks may occur for particular fibres and in particular regions before 100 per cent aggregate utilisation is achieved. Aggregates conceal particular fibre and regional differences. For instance, the non-cellulosic utilisation figures do not show the tendency of nylon capacity utilisation during the period to vary inversely with utilisation of polyester capacity. Capacity utilisation figures are broken down in Table 5.1 in terms of the main types of fibres, but even this breakdown still involves considerable aggregation.

5.2 CAPACITY UTILISATION RELATED TO WITHIN-YEAR PRICE VARIABILITY

Let us first consider whether within-year variations of natural fibre prices are related to capacity utilisation in the man-made fibre industry and then consider between-year price variations. *A priori*, one expects within-year relative price variations of

---

[2] Capacity utilisation was measured by comparing actual production with the estimated production capacity in the man-made fibre industry. Production capacity "is measured at March of the year following production as this measure is the most accurate estimate available of capacity for the calendar year of production" [5]. Wool prices are measured by the price of 21μ clean combing wool at Australian auctions and cotton prices by the price of Memphis Territory SM 1-1/16" CIF Liverpool.

[3] Calculations undertaken since the original analysis show that capacity utilisation in the three broad groups, total cellulose, total non-cellulose and total fibre, fell between 1974 and 1976. Actual percentage figures for 1975 and 1976 are:

Total cellulose fibre:	1975 (73.9),	1976 (80.4)
Total non-cellulose fibre:	1975 (68.6),	1976 (73.0)
Total fibre:	1975 (70.0),	1976 (74.9).

TABLE 5.1

World Capacity Utilisation[(a)] of Man-Made Fibres 1960 to 1974 (Percentages)

FIBRE	1960	1961	1962	1963	1964	1965	1966	1967	1968	1969	1970	1971	1972	1973	1974
Cellulosic yarn															
Rayon	72.4	74.1	80.1	76.7	87.7	82.8	83.3	84.6	87.0	88.5	87.2	92.2	91.9	92.5	86.8
Acetate	74.1	78.5	87.1	85.7	92.5	90.2	91.0	89.4	89.8	87.7	87.2	85.7	82.2	88.5	78.4
Rayon (industrial)	101.5	101.1	112.2	110.4	89.5	93.3	90.0	80.6	85.0	85.0	84.5	89.3	86.2	90.5	84.2
Total	81.0	82.4	90.3	87.3	89.4	87.9	87.3	84.5	87.1	87.2	86.3	89.3	87.2	90.6	83.4
Cellulosic staple															
Rayon	72.1	75.7	89.1	81.9	82.0	86.1	83.8	83.6	84.9	84.6	83.1	84.8	92.7	89.9	86.3
Acetate	54.8	47.3	42.9	58.9	79.3	68.9	72.7	72.4	79.1	66.7	66.6	84.0	76.9	76.0	71.4
Total	71.4	74.4	86.9	81.1	81.9	85.7	83.6	83.4	84.8	85.8	82.9	84.7	89.1	89.7	86.2
Total cellulosic	75.2	77.6	88.3	83.5	84.8	86.5	85.1	83.9	85.7	86.4	84.3	86.5	88.4	90.1	85.1
Non-cellulosic yarn															
Acrylic		-	-	28.6	33.3	37.5	33.3	50.0	66.7	66.7	66.7	71.4	71.4	83.3	100.0
Nylon	79.6	73.3	78.6	80.6	78.1	73.4	75.6	73.5	77.9	77.9	75.9	81.5	82.3	84.6	77.9
Polyester	72.0	77.4	71.4	73.8	80.5	69.5	62.5	65.9	72.8	73.3	68.4	73.1	73.9	81.2	74.2
Other	40.0	38.6	46.2	47.3	62.5	57.5	60.6	55.0	58.3	68.9	73.3	75.0	65.5	84.9	75.0
Total	75.0	70.4	75.0	76.5	76.9	71.4	72.3	70.4	76.2	79.8	73.6	78.5	78.8	83.2	76.3
Non-cellulosic staple															
Acrylic	54.5	55.3	61.7	72.1	70.0	75.5	71.1	68.7	77.6	73.3	76.8	78.3	78.5	87.0	73.7
Polyester	65.9	62.4	70.8	70.6	67.6	67.1	63.4	63.7	71.5	71.9	72.2	70.6	79.1	82.8	75.4
Nylon	88.9	109.4	98.5	86.6	81.4	83.8	88.7	81.1	79.8	84.2	75.4	71.0	77.9	82.1	70.1
Other	64.1	74.3	79.0	70.8	78.0	64.1	70.2	71.6	69.2	72.1	82.8	75.2	71.5	83.2	69.1
Total	63.3	65.9	71.1	73.2	71.4	72.0	69.5	68.0	74.4	71.0	74.8	76.0	78.4	84.4	73.9
Total non-cellulosic	69.8	68.5	73.3	75.1	74.4	71.7	71.0	69.2	75.4	75.4	74.2	77.3	78.6	83.3	75.1
TOTAL MAN-MADE FIBRES	74.0	75.2	83.6	80.8	81.0	80.2	78.4	76.4	80.0	80.9	79.3	81.9	81.8	85.3	78.0

Notes: (a) Capacity utilisation was measured by comparing actual production with the estimated production capacity.[(b)]

(b) Production capacity for the years 1960 to 1974 was measured at March of the year following production, i.e., March 1974 production capacity was used to represent the calendar year 1973.

Source: *Textile Organon* [5].

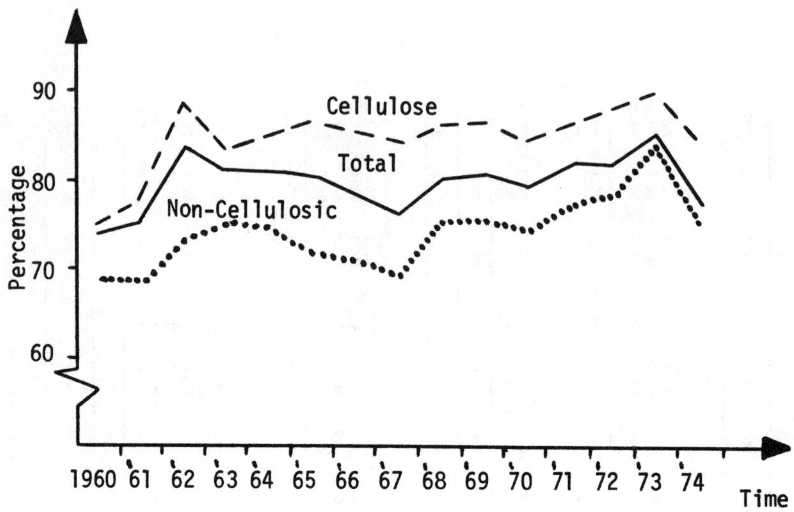

Fig. 5.1

World Capacity Utilisation by Man-Made Fibres 1960 to 1974

Source: Based on Table 5.1.

natural fibres to be subject to considerable random disturbance and therefore, not to be closely related to capacity utilisation of man-made fibres. Nevertheless, this must be specifically examined.

To do this, the within-year coefficients of variation of monthly natural fibre prices (as already depicted in Table 4.1) are compared with man-made fibre capacity utilisation using linear regression analysis. We compare these coefficients of price variation with man-made staple utilisation because man-made staple, rather than yarn, shows greater substitutability with wool and cotton.[4] Competition from non-cellulosic staple is most important. However, rayon (a cellulosic) shows considerable substitutability with cotton.

The linear coefficients of correlation (R) and coefficients of determination (R^2) of within-year natural fibre price variability and capacity utilisation in the production of staples in the man-made fibre industry are set out in Table 5.2. In all cases, a positive relationship exists between the coefficient of variation of natural fibre prices within years and utilisation of man-made

[4] For more information on this see Roberts [4, pp. 20-40, 71-77] and Hirsh and Ellis [2, pp. 108-109].

TABLE 5.2

Coefficients of Linear Correlation and Determination for Within-Year Coefficients of Variation of Monthly Natural Fibre Prices and Capacity Utilisation of Selected Man-Made Fibres 1960 to 1974

Natural fibre for which coefficient of price variation is considered	Type of man-made fibre for which capacity utilisation is considered	Coefficient of correlation R	Coefficient of determination R^2
Wool	Acrylic Staple	0.287	0.082
Wool	Polyester Staple	0.600	0.360
Wool	Nylon Staple	0.334	0.112
Wool	*Total Non-Cellulose Staple*	0.482	0.232
Wool	Rayon Staple	0.371	0.138
Cotton	Acrylic Staple	0.536	0.286
Cotton	Polyester Staple	0.695	0.483
Cotton	*Total Non-Cellulose Staple*	0.703	0.495
Cotton	Rayon	0.447	0.200

Source: Based on Tables 4.1 and 5.1.

staple capacity. While the degree of explanation as measured by R^2 may be low, almost 50 per cent of the variability of the cotton coefficient is "explained" by the utilisation of non-cellulosic staple capacity and almost 25 per cent of the variability of the wool coefficient can be "explained" in this way. The degree of explanation is surprising in view of the likely large influence of random factors in the very short-term. A higher degree of explanation may be expected for between-year variations of natural fibre prices.

5.3 CAPACITY UTILISATION OF MAN-MADE FIBRES RELATED TO BETWEEN-YEAR PRICE VARIABILITY OF NATURAL FIBRES

On the basis of our earlier theories[5] between-year price variability of natural fibres can be expected to rise with the utilisation of available production capacity of competing man-made fibres. To test this relationship, let us measure the relative annual variations of a natural fibre's price by the following identity:

[5] See Chapters 3 and 4.

$$\varepsilon_t = \left| \frac{P_{t+1} - P_t}{P_t} \right| \%$$

where t represents a year, P_t is the average price of the natural fibre (wool or cotton) for a year and P_{t+1} is its average for the succeeding year. Since this is an elasticity-like measure of variation and is therefore in relative terms, it eliminates comparison problems arising from changes in the value of money

The ε-values for wool and cotton for 1960-1974 are shown in Table A5.1, in the Appendix to this Chapter. These values, which show considerable variation, during the period, are compared with capacity utilisation in the man-made fibre industry.

The difficulty arises, however, of deciding on the date at which capacity utilisation should be compared with the ε-values. The elasticity of annual price variation, ε, may, for instance, either be compared with capacity utilisation in the man-made fibre industry in t+1 or t because the time span involved in calculating each ε-value straddles two years. Comparisons are made in this Chapter on both bases but as we shall see, a more significant relationship exists between ε-values and capacity utilisation in t+1 than in t. Table 5.3 sets out the linear correlation coefficients and coefficients of determination for wool ε-values and cotton ε-values in relation to capacity utilisation (in t+1) in the man-made fibre industry during the period 1960-1974. The Table also specifies whether the correlation coefficients are significant at the five per cent level according to the 't' test and lists the linear regression equations relating ε-values to production-capacity utilisation of various types of man-made fibres.

A significant positive linear relationship exists between the elasticity of annual price variation of wool and cotton and the utilisation of total (aggregate) man-made fibre capacity. Almost 35 per cent of the annual relative price variation of wool (R^2 = 0.349) is explained by the utilisation of total man-made fibre capacity (in t+1) and in the case of cotton just over 30 per cent (R^2 = 0.306). However, once we disaggregate the data and consider man-made fibres which show considerable substitutability with wool and cotton, the relationships become more significant. For instance, 70 per cent (R^2 = 0.704) of the elasticity of annual price variation of wool is explained by total non-cellulosic capacity utilisation and 54 per cent (R^2 = 0.538) of the elasticity of annual price variation of cotton is explained in this way. When non-cellulosic staple capacity is considered the degree of explanation for wool rises to 77 per cent (R^2 = 0.767) and that for cotton remains virtually unchanged at 53 per cent (R^2 = 0.527). Non-cellulosic capacity utilisation, especially of staple capacity, significantly explains the relative annual variation of wool and cotton prices. As a scanning of Table 5.3 indicates, the linear correlation coefficients are highest for non-cellulosic

TABLE 5.3

Between-Year Relative Variability of Natural Fibre Prices Related to Man-Made Fibre Capacity Utilisation in t+1 1960 to 1974

Elasticity of Annual Price Variation related to Capacity Utilisation of Man-Made Fibre listed below	Coefficient of Determination R^2	Coefficient of Correlation R	Significant at 5% Level by "t" test	Regression Equation X represents Capacity Utilisation
WOOL ε				
TOTAL MAN-MADE FIBRE	0.349	0.590	Yes	ε = -433+5.7X
Total Non-cellulose	0.704	0.839	Yes	ε = -393+5.5X
Total Non-cellulose Yarn	0.554	0.744	Yes	ε = -289+4.1X
Acrylic Yarn	0.291	0.542	Yes	ε = -15.8+0.65X
Polyester Yarn	0.162	0.402	No	ε = -127+2.0X
Nylon Yarn	0.468	0.684	Yes	ε = -373+5.0X
Total Non-cellulose Staple	0.767	0.876	Yes	ε = -340+5.9X
Acrylic Staple	0.416	0.645	Yes	ε = -138+2.2X
Polyester Staple	0.690	0.830	Yes	ε = -245+3.7X
Nylon Staple	0.100	-0.312	No	ε = 85 -0.78X
Total Cellulose	0.352	0.593	Yes	ε = -426+5.2X
Total Cellulose Yarn	0.109	0.330	No	ε = -247+3.1X
Rayon	0.314	0.560	Yes	ε = -201+2.6X
Acetate Yarn	0.026	-0.161	No	ε = 101-0.95X
Total Cellulose Staple	0.398	0.631	Yes	ε = -344+4.3X
Rayon Staple	0.420	0.648	Yes	ε = -328+4.1X
Acetate Staple	0.103	0.322	No	ε = -28+0.7X
COTTON ε				
TOTAL MAN-MADE FIBRE	0.306	0.553	Yes	ε = -3.7+4.1X
Total Non-cellulose	0.538	0.733	Yes	ε = -270+3.8X
Total Non-cellulose Yarn	0.456	0.675	Yes	ε = -271+3.7X
Acrylic Yarn	0.163	0.404	No	ε = -11+0.39X
Polyester Yarn	0.214	0.462	No	ε = -121+1.8X
Nylon Yarn	0.355	0.596	Yes	ε = -257+3.4X
Total Non-cellulose Staple	0.527	0.726	Yes	ε = -222+3.2X
Acrylic Staple	0.346	0.588	Yes	ε = -101+1.5X
Polyester Staple	0.104	0.323	No	ε = -34+0.66X
Nylon Staple	0.010	-0.100	No	ε = 26-0.19X
Total Cellulose	0.230	0.478	No	ε = -269+3.3X
Total Cellulose Yarn	0.114	0.338	No	ε = -203+2.5X
Rayon Yarn	0.215	0.463	No	ε = -132+1.7X
Acetate Yarn	0.015	0.122	No	ε = -39+0.57X
Total Cellulose Staple	0.188	0.433	No	ε = -184+2.3X
Rayon Staple	0.125	0.354	No	ε = -138+1.7X
Acetate Staple	0.064	0.253	No	ε = -19+0.43X

Source: Based on Tables 5.1 and A5.1.

total capacity and non-cellulose staple capacity. This result is significant, and adds considerable weight to our hypothesis because, as already mentioned, these fibres (especially non-cellulose staple) have shown the greatest substitutability with wool and cotton.

The scatter of between-year relative variability of wool and cotton prices and production-capacity utilisation of non-cellulose staple for the period 1960-1974 as well as the corresponding linear regression lines are shown in Figs 5.2 and 5.3. The linear regression line for wool price variability indicates that between-year relative variability of wool prices can be expected to rise by 4.9 per cent for each one per cent increase in the utilisation of non-cellulose staple capacity. For cotton, a corresponding increase of 3.2 per cent can be expected.

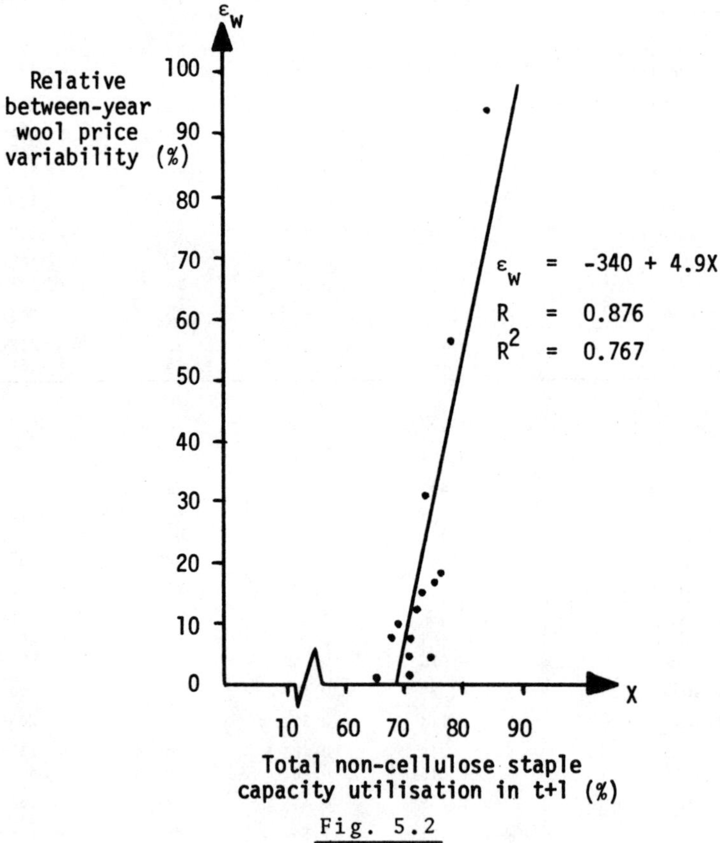

Fig. 5.2

Annual Relative Wool Price Variability and Total Non-Cellulose Staple Capacity Utilisation 1960 to 1974

Source: Based on Tables 5.1, 5.3 and A5.1.

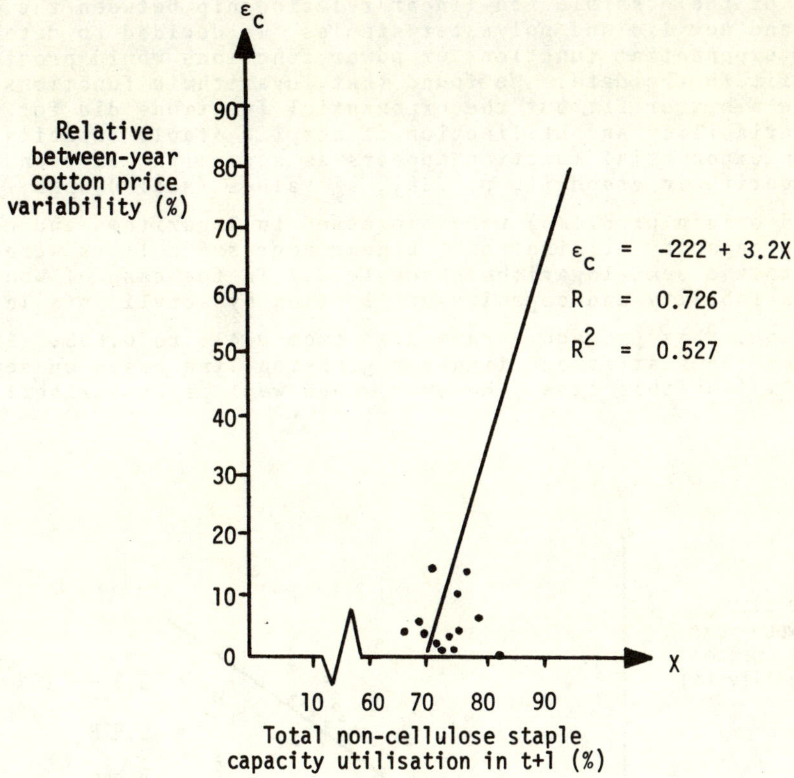

Fig. 5.3

Annual Relative Cotton Price Variability and Total Non-Cellulose Staple Capacity Utilisation 1960 to 1974

Source: Based on Tables 5.1, 5.3 and A5.1.

While the above results do appear to be significant, it might be observed that the close linear relationships between utilisation of the available capacity for the production of non-cellulose staple and the variability of annual wool and cotton prices, as shown in Figs 5.2 and 5.3, arise partly as a consequence of aggregation. For instance, the linear relationship for wool may be improved by the aggregation of a perverse nylon staple relationship[6] with acrylic-polyester-staple relationships[7] in which relative price variability appears to increase at an increasing rate with capacity utilisation.

[6] See Fig. A5.10 in the Appendix to this Chapter.
[7] See Figs A5.6 and A5.8 in the Appendix to this Chapter.

In view of the possible non-linear relationship between the ε-values and acrylic and polyester staples, we decided to determine whether exponential functions or power functions would provide a better fit to the data. We found that logarithmic functions did not give a better fit but the exponential functions did for wool price variability and utilisation of acrylic staple capacity. Since an exponential function appears as a straight line on a semi-logarithmic graph [1, p. 228], ε_w values (scaled up by 10 to avoid origin problems) were expressed in logarithms and compared to capacity utilisation. Linear regression lines were fitted to the semi-logarithmic scatters. In the case of wool price variability and capacity utilisation by acrylic staple production, this procedure raised R^2 from 0.416 to 0.685. Fig. 5.4 shows the scatter and linear regression line based on semi-logarithms for this case. However, when wool price variability

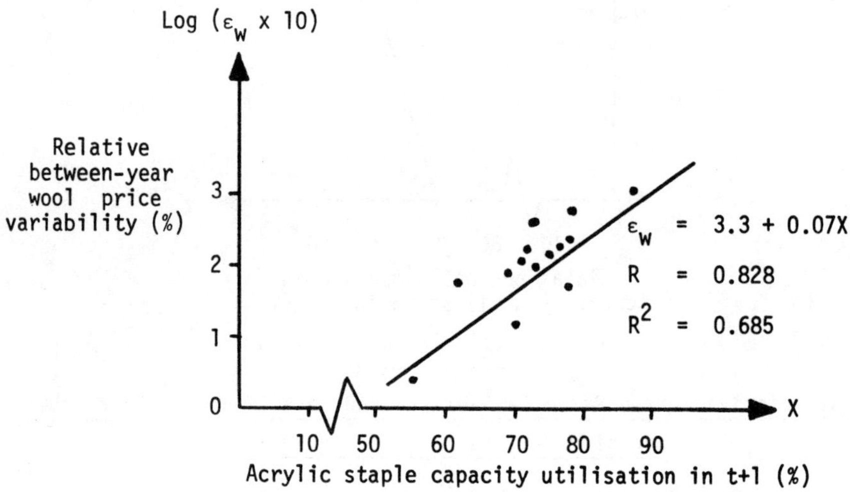

Fig. 5.4

Annual Relative Wool Price Variability and Acrylic Staple Capacity Utilisation 1960 to 1974

Source: Based on Tables 5.1, 5.3 and A5.1.

and utilisation of polyester staple capacity were compared this procedure led to a small reduction in R^2. When applied to cotton this procedure reduced R^2 in relation to capacity utilisation by acrylic staple and increased it in relation to capacity utilisation by polyester staple, but not sufficiently to make R^2 significant.

However, we can conclude that relative between-year variations of wool and cotton prices are positively related to capacity utilisation in the t+1 in various sections of the man-made fibre industry. The relationship is strong in relation to production capacity utilisation by non-cellulosic staples.

The relationship between ε and capacity utilisation in year t rather than t+1 is now considered. In comparison with the relationships in t+1 those in t are found to be less close. Table 5.4 sets out the linear correlation coefficients and coefficients of determination for wool ε-values and cotton ε-values in relation to capacity utilisation in year t in the man-made fibre industry during the period 1960-1974. The Table also indicates whether these correlation coefficients are significant at the five per cent level according to the 't' test and gives the linear regression equations relating ε-values to the production-capacity utilisation (in t) of various man-made fibres.

From Table 5.4 it is clear that the correlation coefficients between ε-values and capacity utilisation in t are, on the whole, much lower than in relation to capacity utilisation in the t+1 (see Table 5.3). In the case of cotton, relationships between ε-values and capacity utilisation by man-made fibres are no longer significant at the five per cent level. In the case of wool, some linear correlation coefficients continue to be significant in relation to capacity utilisation by man-made fibres in t.

Significant linear relationships exist between ε-values for wool and capacity utilisation in t by rayon yarn (R^2 = 0.426), by polyester staple (R^2 = 0.395), by total non-cellulose (R^2 = 0.303) and by nylon yarn (R^2 = 0.290). However, the degree of explanation is lower than in relation to capacity utilisation in t+1. For instance, the linear relationship between capacity utilisation in t by non-cellulose staple explains 37 per cent of the variation of relative wool price whereas 77 per cent was explained by the linear regression on capacity utilisation in t+1. Furthermore, a significant relationship no longer appears to hold for utilisation by acrylic staple when an ordinary linear relationship with ε-values is considered. However, if ε-values are converted to logarithms, the linear regression line based on these is significant for capacity utilisation by acrylic staple in t. On this semi-logarithmic basis, R^2 = 0.450 so that 45 per cent of relative wool price variability is explained in terms of capacity utilisation by acrylic staple. The scatter and linear regression lines for this case are shown in Fig. 5.5. A strong exponential relationship appears to exist between wool ε-values and capacity utilisation in acrylic staple production. This applies to capacity utilisation in t and t+1.

The scatter observations for ε-values for wool and cotton and capacity utilisation by polyester staple and acrylic staple in t are shown in Figs A5.12 and A5.15 in the Appendix to this Chapter. The corresponding regression lines are also drawn in

TABLE 5.4

Between-Year Relative Variability of Natural Fibre Prices Related to Man-Made Fibre Capacity Utilisation in t 1960 to 1974

Elasticity of Annual Price Variation related to Capacity Utilisation of Man-Made Fibre listed below	Coefficient of Determination R^2	Coefficient of Correlation R	Significant at 5% Level by "t" test	Regression Equation X represents Capacity Utilisation
WOOL ε				
TOTAL MAN MADE FIBRE	0.165	0.406	No	ε = -249+3.4X
Total Non-cellulose	0.358	0.598	Yes	ε = -260+4.0X
Total Non-cellulose yarn	0.256	0.507	No	ε = -252+0.26X
Acrylic Yarn	0.323	0.569	Yes	ε = -22.5+0.8X
Polyester Yarn	0.039	0.197	No	ε = -52+1.0X
Nylon Yarn	0.290	0.539	Yes	ε = -288+3.9X
Total Non-cellulose Staple	0.367	0.605	Yes	ε = -221+3.3X
Acrylic Staple	0.233	0.483	No	ε = -78+1.4X
Polyester Staple	0.395	0.628	Yes	ε = -177+2.8X
Nylon Staple	0.207	-0.455	No	ε = 122-1.2X
Total Cellulose	0.244	0.494	No	ε = -249+3.2X
Total Cellulose Yarn	0.102	0.320	No	ε = -214+2.7X
Rayon Yarn	0.426	0.653	Yes	ε = -197+2.6X
Acetate Yarn	0.012	-0.108	No	ε = 67-0.5X
Total Cellulose Staple	0.278	0.527	No	ε = -204+2.7X
Rayon Staple	0.303	0.551	Yes	ε = -165+2.2X
Acetate Staple	0.196	0.443	No	ε = -42.1+0.9X
COTTON ε				
TOTAL MAN-MADE FIBRE	0.02	0.140	No	ε = -62+0.9X
Total Non-Cellulose	0.08	0.290	No	ε = -96+1.4X
Total Non-cellulose Yarn	0.04	0.208	No	ε = -77+1.2X
Acrylic Yarn	0.119	0.345	No	ε = -9.7+0.4X
Polyester Yarn	0.0028	-0.053	No	ε = 25.6-0.21X
Nylon Yarn	0.080	0.284	No	ε = 117+1.6X
Total Non-cellulose Staple	0.112	0.335	No	ε = -95+1.4X
Acrylic Staple	0.074	0.271	No	ε = -32+0.6X
Polyester Staple	0.190	0.436	No	ε = -96+1.5X
Nylon Staple	0.100	-0.310	No	ε = 65-0.6X
Total Cellulose	0.071	0.268	No	ε = -103+1.3
Total Cellulose Yarn	0.0005	0.0235	No	ε = -3.1+0.15X
Rayon Yarn	0.148	0.384	No	ε = -89.1+1.2X
Acetate Yarn	0.036	-0.190	No	ε = 75.6-0.75X
Total Cellulose Staple	0.124	0.353	No	ε = -106+1.4X
Rayon Staple	0.236	0.486	No	ε = -145+1.8X
Acetate Staple	0.080	0.282	No	ε = -20.1+0.46X

Source: Based on Tables 5.1 and A5.1.

and the R^2 and R values are given. Semi-logarithmic and logarithmic transformations of these observations were considered. Linear regression based on logarithmic transformations did not increase R^2 and those based on the semi-logarithmic transformation only increased R^2 in the case of acrylic staple.

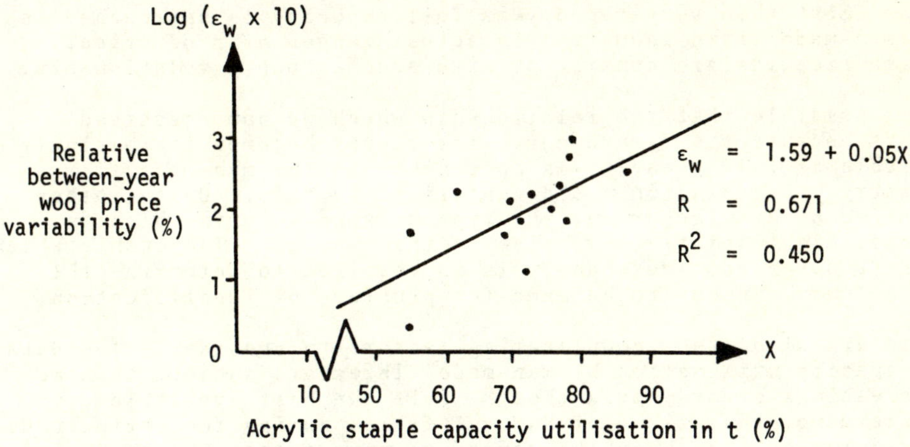

Fig. 5.5

Annual Relative Wool Price Variability and Acrylic Staple Capacity Utilisation 1960 to 1974

Source: Based on Tables 5.1, 5.4 and A5.1.

The above evidence indicates that between-year relative price variability of wool rises with capacity utilisation in various sections of the man-made fibre industry and that this relationship is closer in relation to capacity utilisation in t+1 than in t. In the case of cotton relationships of this type are significant in relation to capacity utilisation in t+1 but not in t. On the whole wool price variability more closely accords with our hypothesis than cotton price variability.[8]

[8] In developing Chapter 6, single regression analysis is again undertaken on wool price variability and capacity utilisation of those man-made fibre categories that yielded significant results in this Chapter. Slight difference in the results reported in Table A6.2 and those in Tables 5.3 and 5.4 occur due to rounding in the original analysis. What is significant however, is the Durbin-Watson tests are applied in the revised exercise and in all categories except one zero auto-correlation is found. The exception is for ε_w and $\log\varepsilon_w$ regressed with total man-made fibre capacity utilisation in t and t+1 where an inconclusive result for auto-correlation is obtained.

5.4 CONCLUSION

While we are encouraged by the evidence in support of our hypothesis that between-year relative price variability of natural fibres rises with increasing capacity utilisation by man-made fibres which are highly substitutable for natural fibres at the margin of use, this does not prove our further hypothesis that the supply curve of fibres is exponential-like (rising very slowly at first then very rapidly as full capacity is approached in the man-made fibre industry) in actual ranges of production. But the results are consistent with such a supply relationship.

It is possible that the relationship which we have observed occurs because the demand curve for fibres becomes increasingly variable as full capacity is approached in the man-made fibre industry. Speculation could contribute to this. On the other hand, both this factor and the type of supply curve which we predict might jointly give rise to the type of relationship which we have observed. More analysis is required to determine this and allowance needs to be made for problems of identification.

There are also other complicating factors to consider. The data on capacity utilisation by man-made fibres are subject to some observations errors, even though it is the best available data. Furthermore, the prices of natural fibres during the period under consideration have been influenced to some extent by government policies. Minimum price support schemes for wool have operated in Australia and for cotton in the U.S.A.[9] It might be argued that these reduced the relative price variability of natural fibre prices, but it is doubtful if this is sufficient to explain the relationship which we observed for the whole range of natural prices.

REFERENCES

[1] ALLEN, R.G.D., *Mathematical Analysis for Economists*, Macmillan, London, 1960.

[2] HIRSH, B.W., and ELLIS, P., *An Introduction to Textile Economics*, Textile Trade Press, Manchester, 1974.

[3] MINFORD, A.P., "Textile Fibre Substitution and Relative Prices", *Australian Journal of Agricultural Economics*, 19, 175-196, 1975.

[4] ROBERTS, R.C., *A Review of Developments in Man-Made Fibres 1960 to 1971*, Wool Economic Research Report No. 24, Bureau of Agricultural Economics, Canberra, 1973.

[5] Textile Economics Bureau Inc., *Textile Organon*, New York, (various June issues).

---

[9] As mentioned by Minford, "the price of cotton in the U.S. was pegged to the user by a government support price scheme and export subsidy system, until 1966, when the government replaced this with a direct production subsidy" [3, p. 178]. To some degree this might explain the smaller correlation values for cotton prices (compared to wool prices) in Tables 5.3 and 5.4.

APPENDIX TO CHAPTER 5

TABLE A5.1

Elasticity of Annual Price Variation ε[(a)] for Wool[(b)] and Cotton[(c)]
1960 to 1974

Year t, t+1	ε for Wool	ε for Cotton
1960, 1961	0.2	3.9
1961, 1962	4.5	1.1
1962, 1963	14.6	2.9
1963, 1964	1.2	0.4
1964, 1965	12.2	0.0
1965, 1966	9.2	3.3
1966, 1967	7.1	5.9
1967, 1968	4.3	9.9
1968, 1969	7.4	14.5
1969, 1970	15.9	4.2
1970, 1971	17.6	13.8
1971, 1972	56.0	6.2
1972, 1973	93.6	78.2
1973, 1974	30.4	0.6

Notes: (a) $\varepsilon_t = \left|\frac{P_{t+1} - P_t}{P_t}\right| \%.$

(b) 21μ clean combing wool at Australian auctions.

(c) Memphis Territory SM 1-1/16" C I F Liverpool.

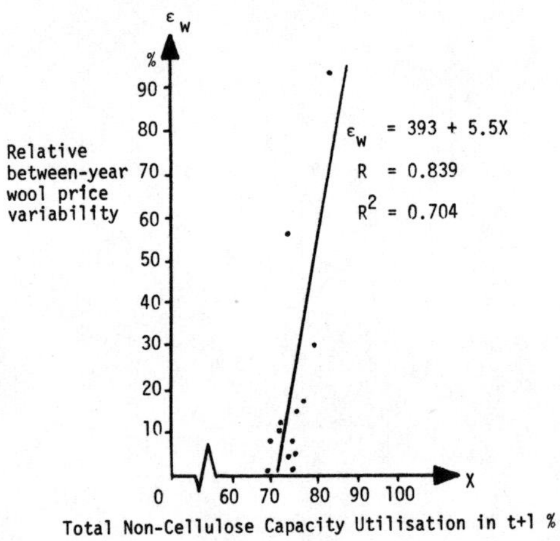

Fig. A5.1

Annual Relative Wool Price Variability and
Total Non-Cellulose Capacity Utilisation 1960-1974

Source: Based on Tables 5.1, 5.3 and A5.1.

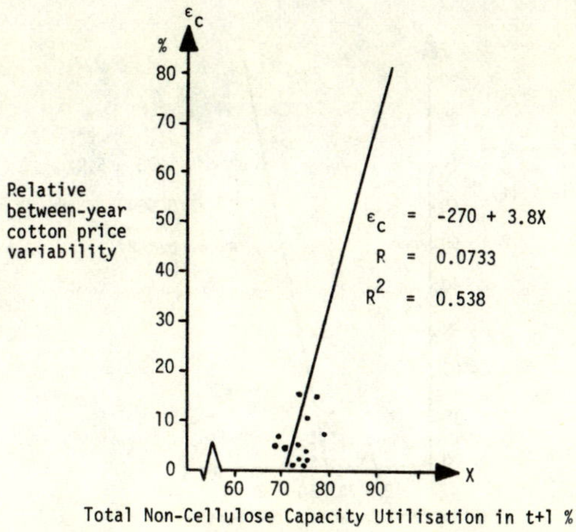

Fig. A5.2

Annual Relative Cotton Price Variability and Total Non-Cellulose Capacity Utilisation 1960-1974

Source: Based on Tables 5.1, 5.3 and A5.1.

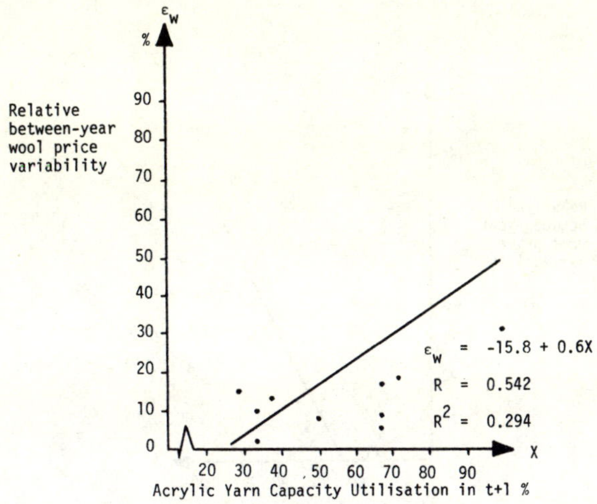

Fig. A5.3

Annual Relative Wool Price Variability and Acrylic Yarn Capacity Utilisation 1960-1974

Source: Based on Tables 5.1, 5.3 and A5.1.

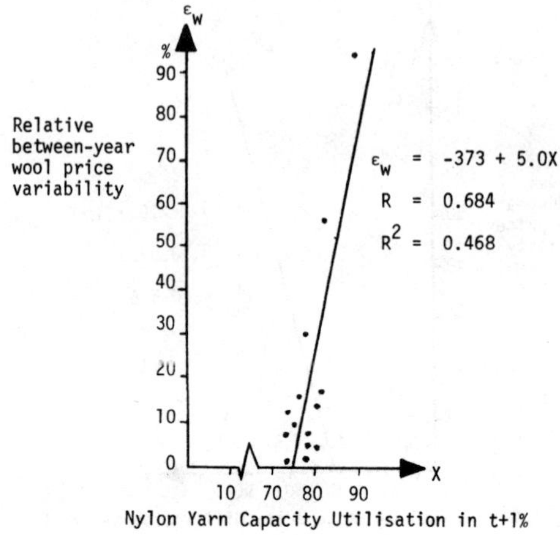

Fig. A5.4

Annual Relative Wool Price Variability and Nylon Yarn Capacity Utilisation 1960-1974

Source: Based on Tables 5.1, 5.3 and A5.1.

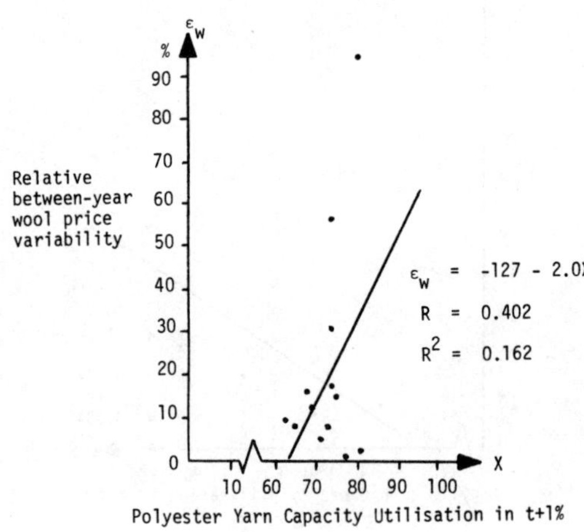

Fig. A5.5

Annual Relative Wool Price Variability and Polyester Yarn Capacity Utilisation 1960-1974

Source: Based on Tables 5.1, 5.3 and A5.1.

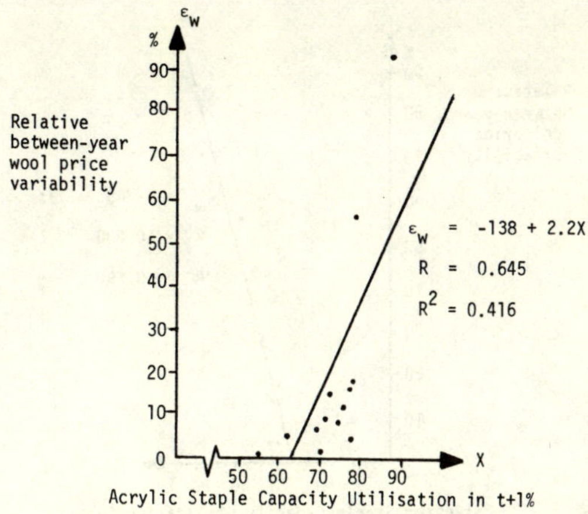

Fig. A5.6

Annual Relative Wool Price Variability and Acrylic Staple Capacity Utilisation 1960-1974

Source: Based on Tables 5.1, 5.3 and A5.1.

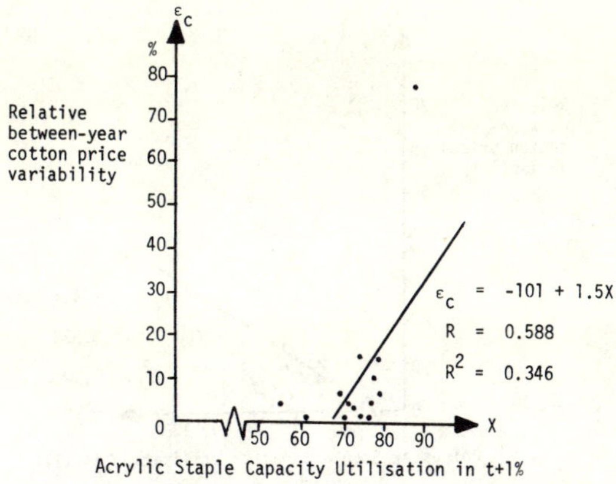

Fig. A5.7

Annual Relative Cotton Price Variability and Acrylic Staple Capacity Utilisation 1960-1974

Source: Based on Tables 5.1, 5.3 and A5.1.

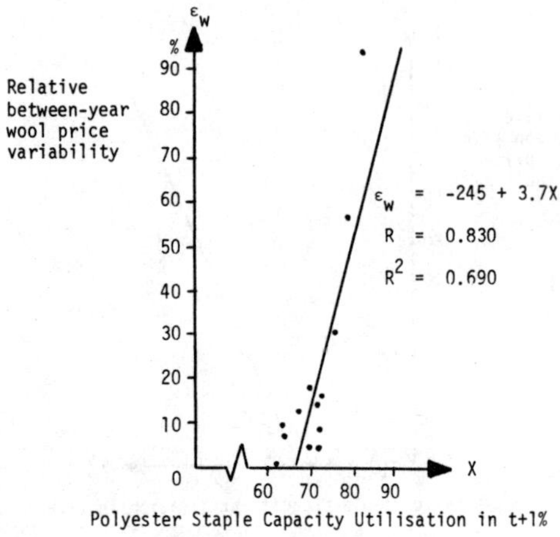

Fig. A5.8

Annual Relative Wool Price Variability and Polyester Staple Capacity Utilisation 1960-1974

Source: Based on Tables 5.1, 5.3 and A5.1.

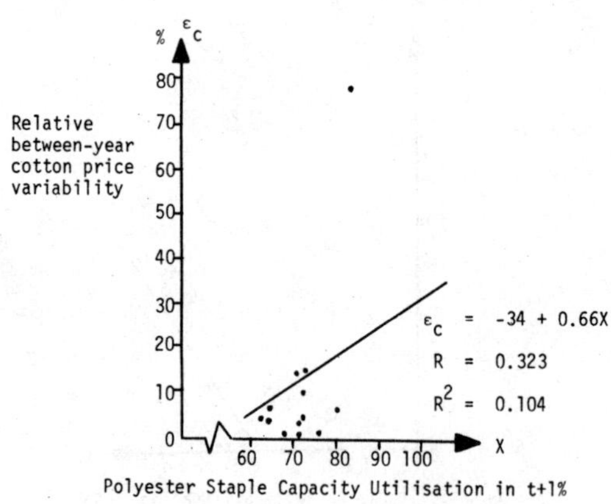

Fig. A5.9

Annual Relative Cotton Price Variability and Polyester Staple Capacity Utilisation 1960-1974

Source: Based on Tables 5.1, 5.3 and A5.1.

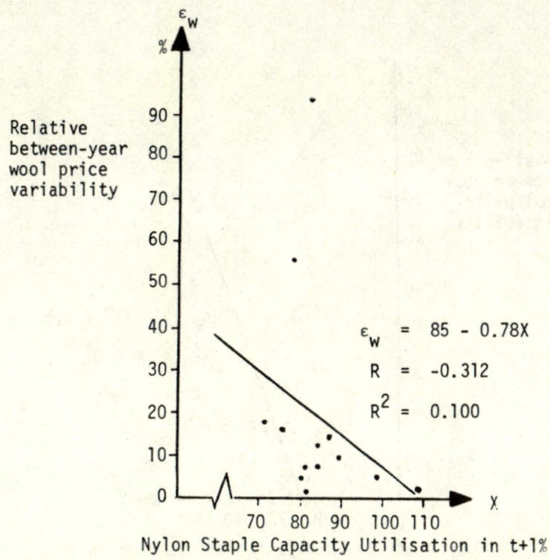

Fig. A5.10

Annual Relative Wool Price Variability and Nylon Staple Capacity Utilisation 1960-1974

Source: Based on Tables 5.1, 5.3 and A5.1.

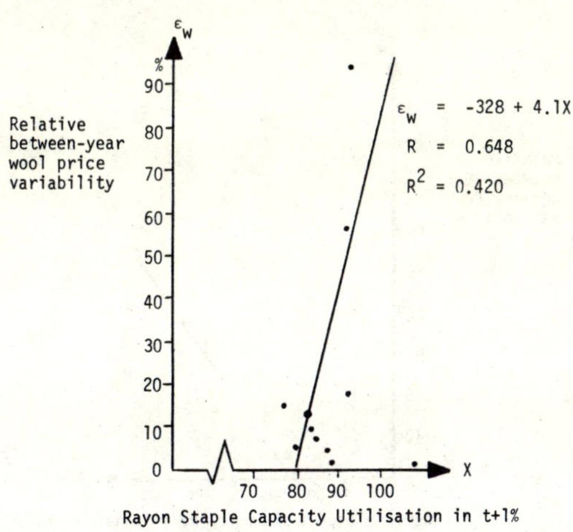

Fig. A5.11

Annual Relative Wool Price Variability and Rayon Staple Capacity Utilisation 1960-1974

Source: Based on Tables 5.1, 5.3 and A5.1.

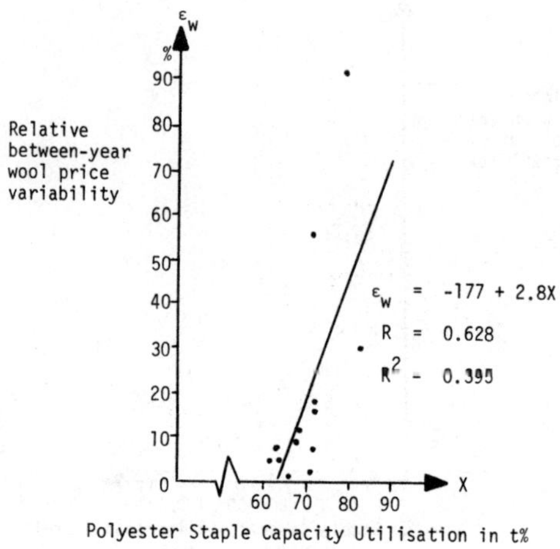

Fig. A5.12

Annual Relative Wool Price Variability and Polyester Staple Capacity Utilisation 1960-1974

Source: Based on Tables 5.1, 5.4 and A5.1.

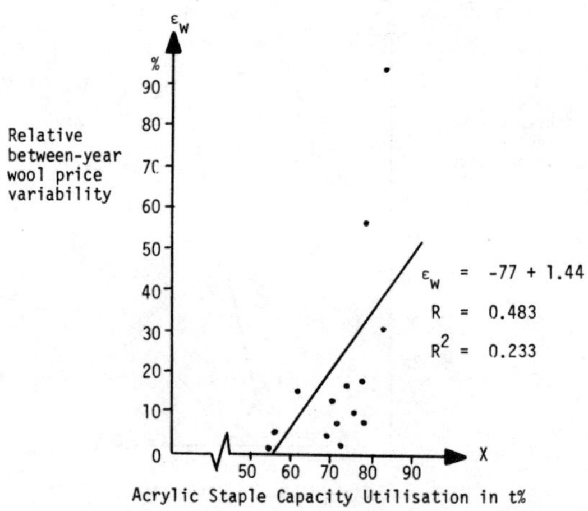

Fig. A5.13

Annual Relative Wool Price Variability and Acrylic Staple Capacity Utilisation 1960-1974

Source: Based on Tables 5.1, 5.4 and A5.1.

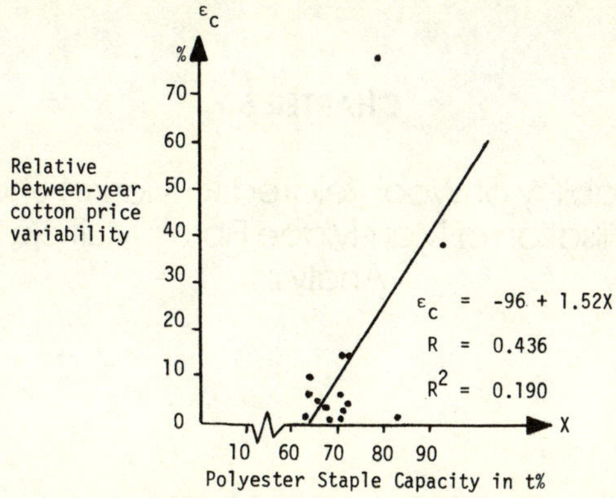

Fig. A5.14

Annual Relative Cotton Price Variability and Polyester Staple Capacity Utilisation 1960-1974

Source: Based on Tables 5.1, 5.4 and A5.1.

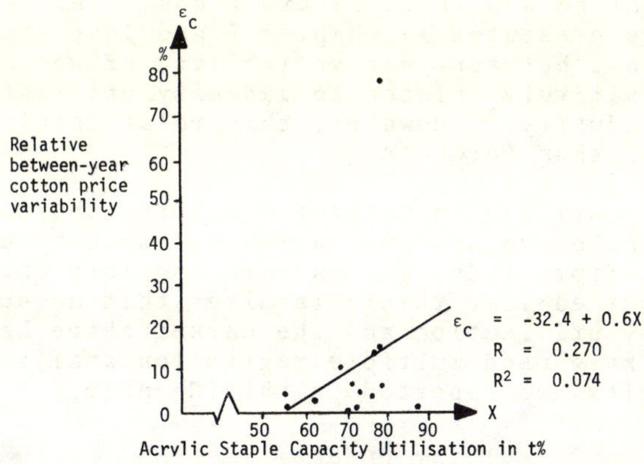

Fig. A5.15

Annual Relative Cotton Price Variability and Acrylic Staple Capacity Utilisation 1960-1974

Source: Based on Tables 5.1, 5.4 and A5.1.

CHAPTER 6

Price Instability of Wool Related to Market Share and Capacity Utilisation of Man-Made Fibres: Multiple Regression Analysis

6.1 INTRODUCTION

In the theoretical model presented in Chapter 3, it was suggested that the relative variability of wool and cotton (natural fibre) prices rises (i) with capacity utilisation in the man-made fibre industry and, (ii) with increases in the share of the fibre market held by man-made fibres. The argument rested principally upon the view that the supply curve of fibres is altered as a result of the growing industrial production of man-made fibres compared to the agricultural supply of natural fibres.

Since in our view capacity utilisation in the man-made fibre industry is likely to be the most important explanatory variable of fluctuations in the relative prices of natural fibres, we first of all considered the empirical relationship between capacity utilisation and these fluctuations. The results of this analysis were presented in Chapter 5 and lend support to the hypothesis that between-year variability of wool and cotton prices is positively related to capacity utilisation in the man-made fibre industry.[1] However, this relationship for wool is much stronger than for cotton.

The analysis outlined in Chapter 5 has the shortcoming of relying only on a single explanatory variable, capacity utilisation in the man-made fibre industry. A more complete empirical consideration of our earlier theory requires that account be taken of both capacity utilisation and the market share held by man-made fibres. We have used multiple regression analysis to do this and the results are reported in this Chapter.

---

[1] The data for capacity utilisation in the man-made fibre industry in the period 1960-1974 is set out in Table 5.1 and the relative variability of wool and cotton prices, ε_w and ε_c, is set out in Table A5.1.

6.2 CAPACITY UTILISATION AND MARKET SHARE OF MAN-MADE FIBRES RELATED TO BETWEEN-YEAR PRICE VARIABILITY OF WOOL

In the earlier analysis, the relative annual variations of a natural fibre's price are measured by

$$\varepsilon_t = \left| \frac{P_{t+1} - P_t}{P_t} \right| \% \qquad (6.1)$$

where t represents a year, P_t is the average price of the natural fibre (wool or cotton) for a year and P_{t+1} is its average for the succeeding year. These values for wool during the period 1960-1974 were regressed against percentage capacity utilisation in the man-made fibre industry in t and in t+1 and the market share held by man-made fibres, in t and t+1. The following linear equation was fitted to the data:

$$\varepsilon_w = a + bX_1 + cX_2 \qquad (6.2)$$

where X_1 is capacity utilisation in the man-made fibre industry (or a sector of it), X_2 is the percentage of fibre market held by man-made fibres (or a selected segment of it) and a, b, and c are parameters. The results are set out in Table 6.1. The results for the semi-logarithmic relationship

$$\text{Log}(\varepsilon_w \times 10) = a + bX_1 + cX_2 \qquad (6.3)$$

which was fitted also to the data, are set out in Table 6.2.

The market share of all man-made fibres and selected types of man-made fibres in total fibre production are set out in Table A6.1 of this Chapter and graphed in Fig. 6.1. In general the multiple regression results set out in Tables 6.1 and 6.2 indicate that capacity utilisation and market share in t+1 show a closer relationship with the independent variables, ε_w and log (ε_w x 10) than do capacity utilisation and market share in t. Possibly log(ε_w x 10) is also more closely related to the independent variables than ε_w but this is not consistently the case.

The estimated linear relationship between ε_w and the independent variables, the percentage utilisation of capacity for producing (all) man-made fibres measured in t+1 and man-made fibre production as a percentage of total fibre production measured in t+1, "explains" two-thirds of the variations of ε_w, (R^2 = 0.665). The estimated linear relationship between log(ε_w x 10) and these independent variables measured either at t or t+1 "explains" two-thirds or more of the variation of log(ε_w x 10), (R^2 = 0.667) in

TABLE 6.1

Between-Year Relative Variability of Wool Price [(a)] (ε_w) [(b)] Related to Capacity Utilisation of Particular Man-Made Fibre Types (X_1) and Particular Man-Made Fibre Types as a % of Total Fibre Production (X_2), in t and t+1 [(c)] 1960 to 1974

Dependent variable: wool price variability in period t	Independent variables: capacity utilisation (X_1) in period t and production as a % of total fibre production (X_2) in period specified below for types of man-made fibre	Mult. coeff. of det. R^2	Mult. coeff. of corr. R	t test on X_1 [(d)] value and result at 5% level of significance		t test on X_2 [(e)] value and result at 5% level of significance		Durbin-Watson test [(f)]: value and result at 5% level of significance		Multiple regression equation: X_1 - represents capacity utilisation; X_2 - represents relative production percentage
ε_w	Total man-made fibre, t	0.364	0.603	1.068	No	1.917	No	1.477	Inc.	$\varepsilon_w = -205+2.15X_1+1.63X_2$
ε_w	Total man-made fibre, t+1	0.665	0.817	2.896	Yes	3.233	Yes	1.764	ZA	$\varepsilon_w = -441+4.85X_1+2.18X_2$
ε_w	Total non-cellulose, t	0.409	0.640	1.112	No	1.129	No	1.705	ZA	$\varepsilon_w = -159+2.19X_1+1.00X_2$
ε_w	Total non-cellulose, t+1	0.739	0.860	3.363	Yes	1.114	No	1.755	ZA	$\varepsilon_w = -336+4.62X_1+0.704X_2$
ε_w	Tot. non-cel. staple, t	0.361	0.601	0.606	No	1.333	No	1.653	ZA	$\varepsilon_w = -83+1.14X_1+2.43X_2$
ε_w	Tot. non-cel. staple, t+1	0.740	0.860	3.500	Yes	0.566	No	1.582	ZA	$\varepsilon_w = -311+4.45X_1+0.738X_2$
ε_w	Acrylic staple, t	0.350	0.522	0.057	No	1.502	No	1.462	Inc.	$\varepsilon_w = -1.99-0.066X_1+8.09X_2$
ε_w	Acrylic staple, t+1	0.518	0.720	1.189	No	1.021	No	1.030	Inc.	$\varepsilon_w = -97+1.370X_1+4.99X_2$
ε_w	Polyester staple, t	0.387	0.622	1.037	No	0.873	No	1.512	Inc.	$\varepsilon_w = -102+1.565X_1+3.55X_2$
ε_w	Polyester staple, t+1	0.709	0.842	3.079	Yes	0.462	No	1.638	ZA	$\varepsilon_w = -222+3.346X_1+1.41X_2$

Notes: (a) 21μ clean combing wool at Australian Auctions.

(b) $\varepsilon_w = \left|\frac{P_{t+1} - P_t}{P_t}\right| \%$

(c) Man-Made Fibre data in t refers to Man-Made Fibre data beginning in 1960 while t+1 refers to the Man-Made Fibre data beginning in 1961.

(d) Critical (Table) values of $t_{.05,11} = \pm 2.201$.

(e) Critical (Table) values of $t_{.05,12} = \pm 2.179$.

(f) Upper and lower range values for the Durbin-Watson Statistic (d) were $d_L = 0.95$, $d_u = 1.54$. ZA = Zero autocorrelation. Inc. = Inconclusive.

Source: Based on data in Tables 5.1, A5.1 and A6.1

TABLE 6.2

Between-Year Relative Variability of Wool Price [a] ($\log \varepsilon_w$)[b] Related to Capacity Utilisation of Particular Man-Made Fibre Types (X_1) and Particular Man-Made Fibre Types as a % of Total Fibre Production (X_2), in t and t+1[c] 1960 to 1974

Dependent variable: relative wool price variability in period t	Independent variables: capacity utilisation (X_1) in period t and production as a % of total fibre production (X_2) in period specified below for types of man-made fibre	Mult. coeff. of det. R^2	Mult. coeff. of corr. R	t test on X_1[d] value and result at 5% level of significance		t test on X_2[e] value and result at 5% level of significance		Durbin-Watson[f] test: value and result at 5% level of significance		Multiple regression equation; X_1 - represents capacity utilisation; X_2 - represents relative production percentage
$\log(\varepsilon_w \times 10)$	Total man-made fibre, t	0.667	0.817	2.292	Yes	3.354	Yes	2.165	ZA	$\log \varepsilon_w = -6.72+0.087X_1+0.054X_2$
$\log(\varepsilon_w \times 10)$	Total man-made fibre, t+1	0.711	0.843	2.450	Yes	4.211	Yes	2.544	ZA	$\log \varepsilon_w = 8.31+0.100X_1+0.069X_2$
$\log(\varepsilon_w \times 10)$	Total non-cellulose, t	0.526	0.726	0.735	No	2.050	No	1.842	ZA	$\log \varepsilon_w = -1.23+0.034X_1+0.042X_2$
$\log(\varepsilon_w \times 10)$	Total non-cellulose, t+1	0.632	0.795	1.385	No	2.169	No	2.115	ZA	$\log \varepsilon_w = -3.16+0.060X_1+0.043X_2$
$\log(\varepsilon_w \times 10)$	Tot. non-cel. staple, t	0.495	0.704	0.111	No	2.316	Yes	1.756	ZA	$\log \varepsilon_w = 0.837+0.005X_1+0.098X_2$
$\log(\varepsilon_w \times 10)$	Tot. non-cel. staple, t+1	0.673	0.820	2.002	No	1.595	No	2.159	ZA	$\log \varepsilon_w = -4.010+0.075X_1+0.061X_2$
$\log(\varepsilon_w \times 10)$	Acrylic staple, t	0.531	0.729	0.775	No	1.429	No	2.094	ZA	$\log \varepsilon_w = -0.001+0.020X_1+0.170X_2$
$\log(\varepsilon_w \times 10)$	Acrylic staple, t+1	0.710	0.843	2.344	Yes	0.956	No	2.132	ZA	$\log \varepsilon_w = -2.36+0.055X_1+0.095X_2$
$\log(\varepsilon_w \times 10)$	Polyester staple, t	0.471	0.687	0.339	No	1.859	No	1.712	ZA	$\log \varepsilon_w = 0.490+0.012X_1+0.184X_2$
$\log(\varepsilon_w \times 10)$	Polyester staple, t+1	0.640	0.800	1.608	No	1.532	No	1.667	ZA	$\log \varepsilon_w = -2.11+0.051X_1+0.136X_2$

Notes: (a) 21µ clean combing wool at Australian Auctions.

(b) $$\log \varepsilon_w = \log(\varepsilon_w \times 10) = \left|\frac{\log(P_{t+1} \times 10) - \log(P_t \times 10)}{\log(P_t \times 10)}\right| \%$$

(c) Man-Made Fibre data in t refers to Man-Made Fibre data beginning in 1960 while t+1 refers to the Man-Made Fibre data beginning in in 1961.

(d) Critical (Table) values of t.05,11 = ±2.201.

(e) Critical (Table) values of t.05,12 = ±2.179.

(f) Upper and lower range values for the Durbin-Watson Statistic (d) were $d_L = 0.95$, $d_u = 1.54$. Z.A. = Zero autocorrelation. Inc. = Inconclusive.

Source: Based on data in Tables 5.1, A5.1 and A6.1.

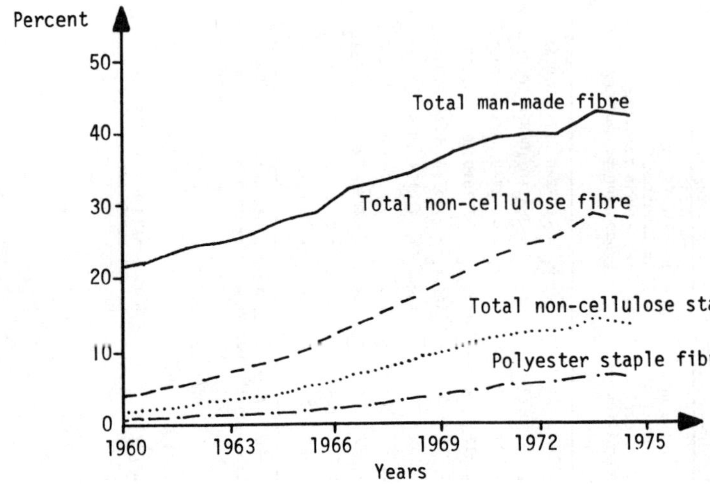

Fig. 6.1

Particular Man-Made Fibre Types as a Percentage of Total Fibre Production 1960 to 1974

Source: Based on Table A6.1.

t and (R^2 = 0.711) in t+1. These results lend support to the hypothesis mentioned at the beginning of this Chapter especially since 't' tests indicate, at the five per cent level of significance, that both capacity utilisation and market share of man-made fibres are significant explanatory variables of the instability of wool prices. Durbin-Watson tests were also applied and indicated zero autocorrelation in each case. The addition of X_2, the market share of man-made fibres, as an explanatory variable adds to the explanation of the instability of wool prices.

Looking at selected segments of the man-made fibre industry (selected on the basis of an apparent high degree of substitutability with wool[2] capacity utilisation in t+1 is found to be a

[2] See Roberts [6, pp. 20-40, 71-77] and Hirsh and Ellis [3, pp. 108-109]. Tests for multicollinearity were completed for the five types of man-made fibres selected. No multicollinearity existed between total man-made fibre capacity utilisation and total man-made fibre share, but some multicollinearity did exist in the other four cases. However, since the introduction of man-made fibre share (as an additional independent variable) had little effect on the multiple R^2 values, the presence of some degree of multicollinearity does not basically affect our conclusions. At this stage, lack of suitably refined data prevents any attempt to isolate the multicollinearity and thereby obtain what might be more significant multiple R^2 values. This is an area in which further research could be undertaken.

TABLE 6.3

Between-Year Relative Variability of Cotton Price[(a)] (ε_c and log ε_c)[(b)] Related to Capacity Utilisation of Particular Man-Made Fibre Types (X_1) and Particular Man-Made Fibre Types as a Percentage of Total Fibre Production (X_2) in t and t+1[(c)] 1960 to 1974

Dependent variable: relative cotton price variability (Y)	Independent variables: capacity utilisation (X_1) in period t and production as a % of total fibre production (X_2) in period specified below for types of man-made fibre	Mult. coeff. of det. R^2	Mult. coeff. of corr. R	t test on X_1[(d)] value and result at 5% level of significance	t test on X_2[(e)] value and result at 5% level of significance	Durbin-Watson[(f)] test: value and result at 5% level of significance	Multiple regression equation: X_1 - represents capacity utilisation; X_2 - represents relative production percentage
ε_c	Total man-made fibre, t	0.137	0.370	0.0868 No	1.275 No	2.671 ZA	$\varepsilon_c = -35+0.158X_1+0.980X_2$
ε_c	Total man-made fibre, t+1	0.473	0.688	2.243 Yes	1.872 No	2.701 ZA	$\varepsilon_c = -325+3.68X_1+1.23X_2$
ε_c	Total non-cellulose, t	0.157	0.396	0.044 No	1.039 No	2.704 ZA	$\varepsilon_c = -9.45+0.08X_1+0.857X_2$
ε_c	Total non-cellulose, t+1	0.546	0.739	2.698 Yes	0.045 No	2.554 ZA	$\varepsilon_c = -274+3.820X_1+0.029X_2$
ε_c	Total non-cell. staple, t	0.155	0.393	-0.266 No	1.208 No	2.615 ZA	$\varepsilon_c = 27.5-0.450X_1+1.968X_2$
ε_c	Total non-cell. staple, t+1	0.577	0.759	2.895 Yes	-0.274 No	2.487 ZA	$\varepsilon_c = -255+3.677X_1-0.357X_2$
$\log(\varepsilon_c \times 10)$	Total man-made fibre, t	0.346	0.588	-1.616 No	2.353 Yes	2.521 ZA	$\log \varepsilon_c = 7.16-0.095X_1+0.058X_2$
$\log(\varepsilon_c \times 10)$	Total man-made fibre, t+1	0.259	0.509	3.364 No	1.856 No	1.776 ZA	$\log \varepsilon_c = -2.32+0.026X_1+0.053X_2$
$\log(\varepsilon_c \times 10)$	Total non-cellulose, t	0.449	0.670	-2.328 Yes	3.119 Yes	2.422 ZA	$\log \varepsilon_c = 9.77-0.127X_1+0.008X_2$
$\log(\varepsilon_c \times 10)$	Total non-cellulose, t+1	0.290	0.538	0.897 No	0.845 No	1.579 ZA	$\log \varepsilon_c = -3.21+0.058X_1+0.025X_2$
$\log(\varepsilon_c \times 10)$	Total non-cell. staple, t	0.391	0.625	-1.870 No	2.770 Yes	2.537 ZA	$\log \varepsilon_c = 7.653-0.099X_1+0.141X_2$
$\log(\varepsilon_c \times 10)$	Total non-cell. staple, t+1	0.288	0.537	0.502 No	1.077 No	1.734 ZA	$\log \varepsilon_c = -1.222+0.030X_1+0.066X_2$

Notes: (a) Memphis Territory SM 1 1/16" CIF Liverpool.

(b) $\varepsilon_c = \left|\frac{P_{t+1} - P_t}{P_t}\right| \%$; $\log \varepsilon_c = \log(\varepsilon_c \times 10) = \left|\frac{\log(P_{t+1} \times 10) - \log(P_t \times 10)}{\log(P_t \times 10)}\right| \%$

(c) Man-Made Fibre data in t refers to Man-Made Fibre data beginning tin 1960 while t+1 refers to the Man-Made Fibre data beginning in 1961.

(d) Critical (Table) values of t.05,11 = ±2.201.

(e) Critical (Table) values of 5.05,12 = ±2.179.

(f) Upper and lower range values for the Durbin-Watson Statistic (d) were $d_L = 0.95$, $d_u = 1.54$.

ZA = Zero autocorrelation. Inc. = Inconclusive.

Source: Based on data in Tables 5.1, A5.1 and A6.1.

significant explanatory variable whereas shares of the total fibre market are not normally significant explanatory variables, according to 't' tests at the five per cent level of significance, of the relative annual variability of wool prices. When capacity utilisation and market share of total non-cellulose, of total non-cellulose staple, of acrylic staple and of polyester staple are measured in t+1 and used as independent variables, the estimated linear relationships for either ε_w or $\log(\varepsilon_w \times 10)$ "explain" two-thirds or more of the variation of these variables, except in the case of ε_w and acrylic capacity utilisation and market share. (See relevant R^2 values in Tables 6.1 and 6.2.) However, capacity utilisation is the only really significant explanatory variable in these cases. According to the 't' test, the variable X_2, the market share of man-made fibres, does not add significantly to the explanation of the relative variability of wool prices, even though it raises R^2 values slightly. The Durbin-Watson test again indicated zero autocorrelation in each case (with the exception of ε_w and acrylic capacity utilisation and market share in t+1).

Table 6.3 sets out our multiple regression results for cotton in which ε_c and $\log(\varepsilon_c \times 10)$ are regressed against man-made fibre capacity utilisation and market share. While the relevant R^2 values are all non-zero they do not indicate a close relationship between the variability of cotton prices and the explanatory variables. This might be a reflection of the influence of U.S. Government policy on cotton prices[3] as well as the larger share of the fibre market held by cotton which makes it a much larger buffer or "shock-absorber" than wool.

In the case of wool, the evidence strongly supports the hypothesis that the variability of wool prices is positively related to the utilisation of production capacity in the man-made fibre industry. In addition, the market share of *combined* man-made fibres in relation to all fibres is a significant explanatory variable of the variability of wool prices. While other factors also may influence the instability of wool prices, the apparent significance of the two variables which we have isolated cannot be dismissed lightly. Our earlier theory provides a possible explanation of why these variables are important.

It is interesting also to observe the results of pair-wise regressions reported in Tables A6.2 and A6.3. Results for the linear regression of ε_w and $\log(\varepsilon_w \times 10)$ in relation to capacity utilisation in various sectors of the man-made fibre industry are given in Table A6.2. Results for the linear regression of ε_w and $\log(\varepsilon_w \times 10)$ in relation to the market share of various man-made fibres are reported in Table A6.3. For nearly all

[3] Chapter 2 (section 2.4) outlines this policy and considers some of its implications.

sectors of the man-made fibre industry considered in these tables, capacity utilisation and market share are significant variables, at the five per cent level when 't' test is applied. Furthermore, as in the multiple regression case the Durbin-Watson statistics given in these tables indicate that in nearly all cases autocorrelation is absent.

The sizes of the coefficients of determination (R^2 values) vary considerably in Tables 6.1, 6.2, A6.2 and A6.3 but several are in excess of 0.7 in one or more tables. The coefficients of determination in Tables 6.1, 6.2, A6.2 and A6.3 are set out in Table 6.4 for each sector in which an R^2 value in any of these tables exceeds 0.7. In these instances, one or more of the linear regressions which we have considered "explain" 70 per cent or more of the variation of ε_w or $\log(\varepsilon_w \times 10)$.

TABLE 6.4

Coefficients of Determination for Regressions Analysed in Tables 6.1, 6.2, A6.2 and A6.3 Where Any R^2 Value Exceeds 0.7

Independent variable	Man-made fibre sector	R^2 in Table A6.2 dependent variable capacity utilisation	R^2 in Table A6.3 dependent variable relative market share	R^2 in Tables 6.1 & 6.2 combined dependent variables
ε_w	Total Non-Cellulose Staple in t+1	0.732	0.451	0.740
ε_w	Total Non-Cellulose in t+1	0.710	0.471	0.739
ε_w	Polyester Staple in t+1	0.703	0.458	0.709
$\log(\varepsilon_w \times 10)$	Total Man-Made Fibre in t+1	0.246	0.554	0.711
$\log(\varepsilon_w \times 10)$	Acrylic Staple in t+1	0.686	0.565	0.710

Source: Based on Tables 6.1, 6.2, A6.2 and A6.3.

Seventy per cent or more of the variation of ε_w is "explained" in the two-variable regression case by utilisation of capacity for producing total non-cellulose staple ($R^2 = 0.732$), for producing total non-cellulose ($R^2 = 0.710$) and for producing polyester staple ($R^2 = 0.703$). Multiple regression using capacity utilisation and market share as independent variables raises these coefficients, slightly, to 0.740, 0.739 and 0.709 respectively. In the case of $\log(\varepsilon_w \times 10)$ only coefficients of *multiple* determination exceed 0.7. This is so in relation to total man-made fibres and acrylic staple. In the case of total man-made fibres the addition of market share as an explanatory variable adds significantly to the size of R^2 but this is not so for acrylic staple where almost 70 per cent of the variation in log (ε_w x 10) is "explained" in the single regression involving capacity utilisation.

6.3 CONCLUSION

It is clear from Table 6.4 and some of our other results that capacity utilisation in various sectors of the man-made fibre industry is closely related to fluctuations in wool prices and that market share of man-made fibres, while not of negligible significance, is of secondary importance in its relationship to these fluctuations. Such results are consistent with our earlier theories and tend to support the view that the growing relative importance of man-made fibres may add to relative fluctuations in the price of wool. In all cases considered above, the regression equations indicate that the variability of wool prices is positively related to capacity utilisation in the man-made fibre industry and also to the market share of man-made fibres. We do not, as other writers apparently do [2], [8], see the growth of the man-made fibre industry as necessarily exerting a stabilising influence on the price of natural fibres.

REFERENCES

[1] Commonwealth Secretariat, *Wool Intelligence*, London, (various issues).

[2] DUANE, P., *Analysis of Wool Price Fluctuations*, Wool Economics Research Report No. 23, Bureau of Agricultural Economics, Canberra, 1973.

[3] HIRSH, B.N., and ELLIS, P., *An Introduction to Textile Economics*, The Textile Trade Press, Manchester, 1974.

[4] International Cotton Advisory Committee, *Cotton - World Statistics*, Washington, (various issues).

[5] International Wool Textile Organisation and International Wool Study Group, *Wool Statistics*, London, (various issues).

[6] ROBERTS, R.C., *A Review of Recent Developments in Man-Made Fibres 1960 to 1971*, Wool Economic Research Report No. 24, Canberra, 1973.

[7] Textile Economics Bureau Inc., *Textile Organon*, New York, (various issues).

[8] TIER, T.J., and KIDMAN, P.R., "Price Movements in and Between the Wool, Wool Tops and Worsted Yarn Spot and Futures Markets", *Quarterly Review of Agricultural Economics*, 24, 63-87, 1974.

APPENDIX TO CHAPTER 6

TABLE A6.1

Total Fibre Production, Production of Particular Man-Made Fibre Types and their Relative Share of Total Fibre Production 1960 to 1974

(Production in M.Kg.)

Year	Total fibre production	Total man-made fibre		Total non-cellulose fibre		Total non-cellulose staple fibre		Acrylic staple fibre		Polyester staple fibre	
		Production	Percentage of total production	Production	Percentage of total production	Production	Percentage of total production	Production	Percentage of total production	Production	Percentage of total production
1960	14,823	3,247	21.9	639	4.3	285	1.9	109	0.7	87	0.6
1961	14,749	3,442	23.3	752	5.1	334	2.3	121	0.8	103	0.7
1962	15,779	3,845	24.4	981	6.2	442	2.8	163	1.0	148	1.0
1963	16,752	4,291	25.6	1,232	7.4	555	3.3	209	1.3	187	1.2
1964	17,706	4,918	27.8	1,632	9.2	713	4.0	298	1.7	231	1.3
1965	18,410	5,314	28.9	1,975	10.7	926	5.0	401	2.2	320	1.8
1966	17,822	5,713	32.1	2,370	13.3	1,134	6.4	455	2.6	427	2.4
1967	18,032	6,047	33.5	2,730	15.2	1,356	7.5	537	3.0	538	3.0
1968	20,352	7,109	34.9	3,578	17.6	1,813	8.9	729	3.6	244	3.7
1969	20,612	7,739	37.5	4,178	20.3	2,142	10.4	849	4.2	892	4.4
1970	20,887	8,126	38.9	4,693	22.5	2,428	11.6	1,021	4.9	1,007	4.9
1971	22,835	9,037	39.6	5,595	24.5	2,845	12.5	1,167	5.2	1,189	5.2
1972	24,690	9,875	40.0	6,342	25.7	3,140	12.7	1,264	5.2	1,380	5.6
1973	26,266	11,232	42.8	7,569	28.8	3,769	14.4	1,562	6.0	1,640	6.3
1974	26,291	11,037	42.0	7,455	28.4	3,683	13.6	1,444	5.5	1,703	6.5

Source: Based on statistics in *Wool Intelligence* [1], *Cotton - World Statistics* [4], *Wool Statistics* [5] and *Textile Organon* [7].

TABLE A6.2

Between-Year Relative Variability of Wool Price[(a)] (ε_w and log ε_w)[(b)] Related to Capacity Utilisation of Particular Man-Made Fibre Types in t and t+1[(c)] 1960 to 1974

Dependent variable: relative wool price variability (Y)	Independent variable: capacity utilisation of types of man-made fibre (X), in period specified below	Coefficient of determination R^2	Coefficient of correlation R	t test[(d)] value and result at 5% level of significance		Durbin-Watson test[(e)] value and result at 5% level of significance		Regression equation (X represents capacity utilisation of particular man-made fibre category)
ε_w	Total man-made fibre, t	0.169	0.411	1.16	No	1.16	Inc.	ε_w = -251+3.4X
ε_w	Total man-made fibre, t+1	0.347	0.589	2.52	Yes	1.02	Inc.	ε_w = -428+5.6X
ε_w	Total non-cellulose, t	0.349	0.590	2.64	Yes	1.79	ZA	ε_w = -261+3.8X
ε_w	Total non-cellulose, t+1	0.710	0.843	5.42	Yes	1.73	ZA	ε_w = -399+5.6X
ε_w	Total non-cell. staple, t	0.266	0.516	2.17	Yes	1.88	ZA	ε_w = -195+2.9X
ε_w	Total non-cell. staple, t+1	0.732	0.856	5.73	Yes	1.66	ZA	ε_w = -343+5.0X
ε_w	Acrylic staple, t	0.228	0.477	1.96	No	1.45	ZA	ε_w = -76+1.4X
ε_w	Acrylic staple, t+1	0.473	0.688	3.28	Yes	0.86	ZA	ε_w = -148+2.3
ε_w	Polyester staple, t	0.349	0.590	2.64	Yes	1.53	ZA	ε_w = -161+2.6X
ε_w	Polyester staple, t+1	0.708	0.836	5.34	Yes	1.63	ZA	ε_w = -243+3.7X
log(ε_w x 10)	Total man-made fibre, t	0.354	0.595	2.67	Yes	1.27	Inc.	log ε_w = -8.2+0.13X
log(ε_w x 10)	Total man-made fibre, t+1	0.246	0.496	1.98	Yes	1.05	Inc.	log ε_w = -7.9+0.12X
log(ε_w x 10)	Total non-cellulose, t	0.366	0.605	2.74	Yes	1.74	ZA	log ε_w = -5.4+0.1X
log(ε_w x 10)	Total non-cellulose, t+1	0.475	0.689	3.29	Yes	1.79	ZA	log ε_w = -7.0+0.12X
log(ε_w x 10)	Total non-cell. staple, t	0.270	0.519	2.19	Yes	1.81	ZA	log ε_w = -3.7+0.08X
log(ε_w x 10)	Total non-cell. staple, t+1	0.597	0.773	4.22	Yes	2.09	ZA	log ε_w = -6.6+0.12X
log(ε_w x 10)	Acrylic staple, t	0.451	0.672	3.27	Yes	2.25	ZA	log ε_w = -1.6+0.05X
log(ε_w x 10)	Acrylic staple, t+1	0.686	0.828	5.12	Yes	2.07	ZA	log ε_w = -3.3+0.07X
log(ε_w x 10)	Polyester staple, t	0.319	0.565	2.47	Yes	2.47	ZA	log ε_w = -2.6+0.06X
log(ε_w x 10)	Polyester staple, t+1	0.563	0.750	3.93	Yes	1.40	ZA	log ε_w = -4.2+0.09X

Notes: (a) 21μ clean combing wool at Australian Auctions.

(b) $\varepsilon_w = \left|\frac{P_{t+1} - P_t}{P_t}\right|\%$; $\log \varepsilon_w = \log(\varepsilon_w \times 10) = \left|\frac{\log(P_{t+1} \times 10) - \log(P_t \times 10)}{\log(P_t \times 10)}\right|\%$

(c) Man-made Fibre data in t refers to Man-Made Fibre data beginning in 1960 while t+1 refers to the Man-Made Fibre data beginning in 1961.

(d) Critical (Table) values of $t_{.05,12}$ = ±2.160

(e) Upper and lower range values for the Durbin-Watson statistic (d) were d_L = 1.08, d_u = 1.36.

ZA = Zero autocorrelation. Inc. = Inconclusive.

Source: Based on data in Tables 5.1 and A5.1. Note that recomputation resulted in very slight differences between the ε_w result above and those of Tables 5.3 and 5.4.

TABLE A6.3

Between-Year Relative Variability of Wool Price[a] (ε_w and log ε_w)[b] Related to Production of Particular Man-Made Fibre Types as a Percentage of Total Fibre Production in t and t+1[c] 1960 to 1974

Dependent variable: relative wool price variability (Y)	Independent variables: production as a % of total fibre production for types of man-made fibre, in period specified below	Coefficient of determination R^2	Coefficient of correlation R	t test[d] value and result at 5% level of significance		Durbin-Watson test[e] value and result at 5% level of significance		Regression equation (X represents man-made fibre production as a percentage of total fibre production
ε_w	Total man-made fibre, t	0.305	0.552	2.39	Yes	1.36	ZA	ε_w = -43.8+1.9X
ε_w	Total man-made fibre, t+1	0.410	0.640	2.89	Yes	1.37	ZA	ε_w = -62.7+2.5X
ε_w	Total non-cellulose, t	0.348	0.590	2.63	Yes	1.41	ZA	ε_w = -7.8+1.7X
ε_w	Total non-cellulose, t+1	0.471	0.687	3.27	Yes	1.49	ZA	ε_w = -15.8+2.1X
ε_w	Total non-cell. staple, t	0.341	0.584	2.59	Yes	1.43	ZA	ε_w = -6.1+3.2X
ε_w	Total non-cell. staple, t+1	0.451	0.072	3.14	Yes	1.39	ZA	ε_w = -13.4+4.0X
ε_w	Acrylic staple, t	0.350	0.591	2.64	Yes	1.45	ZA	ε_w = -5.9+7.8X
ε_w	Acrylic staple, t+1	0.457	0.676	3.18	Yes	1.36	ZA	ε_w = -12.8+9.6X
ε_w	Polyester staple, t	0.333	0.577	2.55	Yes	1.39	ZA	ε_w = -2.8+6.8X
ε_w	Polyester staple, t+1	0.458	0.677	3.19	Yes	1.50	ZA	ε_w = -9.3+8.4X
log(ε_w x 10)	Total man-made fibre, t	0.516	0.718	3.72	Yes	1.75	ZA	log ε_w = -0.77+0.065X
log(ε_w x 10)	Total man-made fibre, t+1	0.553	0.744	3.86	Yes	1.90	ZA	log ε_w = -0.55+0.74X
log(ε_w x 10)	Total non-cellulose, t	0.510	0.714	3.68	Yes	1.74	ZA	log ε_w = 1.12+0.05X
log(ε_w x 10)	Total non-cellulose, t+1	0.568	0.754	3.97	Yes	1.91	ZA	log ε_w = 0.931+0.06X
log(ε_w x 10)	Total non-cell. staple, t	0.495	0.703	3.57	Yes	1.73	ZA	log ε_w = 1.16+0.10X
log(ε_w x 10)	Total non-cell. staple, t+1	0.554	0.744	3.86	Yes	1.82	ZA	log ε_w = 1.00+0.12X
log(ε_w x 10	Acrylic staple, t	0.510	0.713	3.66	Yes	1.79	ZA	log ε_w = 1.17+0.25X
log(ε_w x 10)	Acrylic staple, t+1	0.565	0.752	3.95	Yes	1.85	ZA	log ε_w = 1.0+0.28X
log(ε_w x 10)	Polyester staple, t	0.466	0.683	3.37	Yes	1.68	ZA	log ε_w = 1.28+0.21X
log(ε_w x 10)	Polyester staple, t+1	0.555	0.745	3.87	Yes	1.79	ZA	log ε_w = 1.1+0.24X

Notes: (a) 21μ clean combing wool at Australian Auctions.

(b) $\varepsilon_w = \left|\frac{P_{t+1} - P_t}{P_t}\right|\%$; $\log \varepsilon_w = \log(\varepsilon_w \times 10) = \left|\frac{\log(P_{t+1} \times 10) - \log(P_t \times 10)}{\log(P_t \times 10)}\right|\%$

(c) Man-Made Fibre data in t refers to Man-Made Fibre data beginning in 1960 while t+1 refers to the Man-Made Fibre data beginning in 1961.

(d) Critical (Table) values of t.05,12 = ±2.160

(e) Upper and lower range values for the Durbin-Watson statistic (d) were d_L = 1.108, d_u = 1.36.

ZA = Zero autocorrelation. Inc. = Inconclusive.

Source: Based on data in Tables 5.1 and A6.1.

CHAPTER 7

Textile Manufacturers' Choice of Fibres, Costs and Reactions to Price Instability

7.1 INTRODUCTION

The purpose of this Chapter is to discuss some of the factors which influence textile producers' choice of fibres and in particular to consider the effect which instability of a fibre's price may have on this choice. As the previous discussion indicates, instability of natural fibre prices is likely to be a continuing problem despite the growing share of the fibre market accounted for by man-made fibres.

The relative stability of the prices of man-made fibres may be one factor tending to entice textile manufacturers into substituting man-made fibres for natural fibres such as wool. Indeed, the Australian Wool Corporation in a recent report [4], suggested that instability of wool prices generates a great deal of uncertainty and diminishes the competitiveness of wool as far as yarn manufacturers, weavers, knitters and makers-up are concerned because all of these groups have to plan for forward commitments and deliveries. But, as discussed below, price instability is only one factor influencing the choice of fibres by textile producers and an attempt will be made to assess its relative importance in the substitution of man-made fibres for natural fibres.

To this end, this Chapter relies heavily on the results of a survey of Australian textile manufacturers conducted by the authors [46]. The results of our survey when supplemented by the findings of previous researchers provides evidence on, and reasons for, fibre substitution between natural and man-made fibres. It also provides evidence on the attitudes of yarn and fabric manufacturers to price instability and in particular indicates whether or not they would prefer stable wool prices. In addition, our survey yielded information about the cost curves of textile manufacturers - information which is interesting from an industrial economics viewpoint, and which has a bearing on the way in which the returns of manufacturers are altered by instability of fibre prices. The last matter is explored in a theoretical section at the end of this Chapter.

We conducted a mail survey of Australian yarn and fabric manufacturers and topmakers in mid-1975 to obtain evidence on the factors mentioned above. An attempt was made to reach all Australian firms and a total of 198 textile manufacturers were sent copies of the *Questionnaire for Australian Yarn and Fabric Manufacturers*.[1] Of these 198, 111 producers replied. Some had, however, ceased operations or were not producers of yarn and fabric and as a result only 60 of the returns were relevant to our research. The response rate was estimated to be of the order of 56 per cent. The actual reply rate (111/198) was also 56 per cent. From Table 7.1 it can be seen that of the 60 relevant returns, the split-up was 19 firms manufacturing yarn, 25 manufacturing fabric and 16 manufacturing yarn and fabric.

TABLE 7.1

Number of Respondent Firms by Manufacturing Process

Manufacturing Firm	Total No.
Yarn	19
Fabric	25
Yarn and Fabric	16
Total Firms	60

Source: First survey.

In addition, a total of 14 firms were sent a copy of the *Questionnaire for Australian Topmakers*[2] and four returns were received. The questions about fibre substitution, price instability and cost relationships were identical to those asked in the *Questionnaire for Australian Yarn and Fabric Manufacturers*, but other questions were modified to make these applicable to topmakers. Only brief remarks are made about the responses of topmakers because only a small number responded and there are few topmakers in Australia.

7.2 SUBSTITUTION BETWEEN NATURAL AND MAN-MADE FIBRES

Fibre Substitution Patterns

Broad patterns: It has already been pointed out in Chapter 1 that man-made fibres have made significant inroads into markets formerly dominated by wool and cotton. This trend began prior to World War II and accelerated from the beginning of the 1960s particularly in the case of synthetics. Synthetics now account

[1] A copy of the *Questionnaire for Australian Yarn and Fabric Manufacturers* is included in the Appendix to this Chapter.

[2] A copy of the *Questionnaire for Australian Topmakers* is also included in the Appendix to this Chapter.

for about 30 per cent of the world's supply of apparel fibres. A recent F.A.O. report [16] noted that the ability of the petrochemical industries to obtain their raw materials at low prices has been fundamental to the competitive strength of synthetics. Despite the rise in crude oil prices, the costs of synthetic yarns has not increased significantly.

Man-made fibres (particularly the synthetics) have penetrated into many end-uses at the expense of natural fibres. These include hand knitting yarn, carpet (especially tufted carpets), apparel fabric (especially double-knit fabric, woven worsted apparel fabric, and woven woollen apparel fabric), sweaters and socks.[3] An example of the considerable impact of synthetics can be illustrated by the case of double-knit fabric. Between 1960 and 1971, the consumption of fibre used in the production of double-knit fabric by the world wool textile industry increased spectacularly (by thirteenfold or 25.3 per cent per annum) and non-cellulosics amounted for over three-quarters of the fibre concerned in this end use in 1971 compared to 22 per cent in 1960 with shares of all other fibre types declining [24].

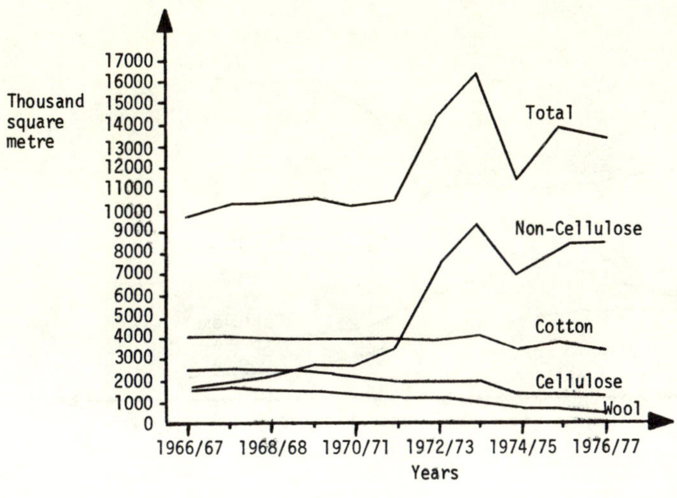

Fig. 7.1

Australian Production of Textiles,[(a)] Monthly Average - Woven Cloth 1966-67 to 1975-76 (Volume - '000 sq. m.)

Note: (a) Each category includes mixtures predominately of the fibre named.

Source: Based on statistics in *Monthly Review of Statistics* [3].

---

3 For a more indepth examination of this penetration see Chapter 1 and reference cited therein.

In his study of the U.S. wool textile industry, Sault found "that fibre substitution was the main cause of the decline in wool consumption in the United States during the period 1949-1967" [37, p. 29]. Similar fibre substitution patterns have been found in other studies [8], [9], in the United Kingdom and the E.E.C., and Pender [32], in a study of fibre substitution in Australia, undertaken in the mid-1960s, also found evidence of considerable fibre substitution between natural and man-made fibre in several end-uses.

This trend towards fibre substitution in Australia has heightened in the years since Pender's study. Evidence of this can be seen from the changing patterns of fibre use in textile production. As an example the increasing influence of synthetic fibre usage in the production of woven cloth in the 11 year period 1966-67 to 1976-77 is illustrated in Figs 7.1 and 7.2; the increase in the volume of use of synthetic fibre in woven cloth production has been dramatic as Fig. 7.1 indicates. In percentage terms, Fig. 7.2 shows that the use of synthetic fibre has risen from 17.1 per cent of total woven cloth production in 1966-67 to 61.0 per cent in 1976-77. This increase has been at the expense of all other fibres.

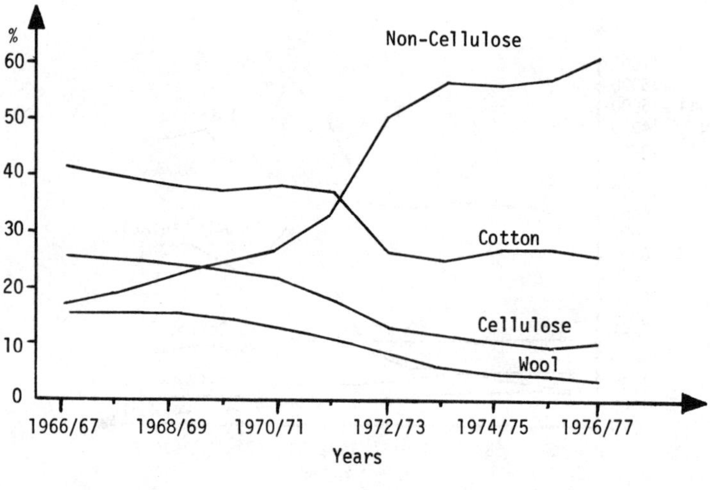

Fig. 7.2

Australian Production of Textiles,[a] Monthly Average - Woven Cloth 1966-67 to 1975-76 (Percentage)

Note: (a) Each category includes mixtures predominately of the fibre named.

Source: Based on statistics in *Monthly Review of Statistics* [3].

Results from Australian survey: In the light of the substitution trends outlined above, we tried to find out from our survey whether individual Australian textile manufacturers could and did alter their fibre mixtures and why.

Of the 35 yarn manufacturers responding to the question on fibre substitution[4] 28 indicated that they could vary fibre mixes and 20 or 57 per cent actually did vary fibre mixes. These 20 yarn producers were classified according to the percentage weight of a fibre in their yarn production.

Four categories of percentage fibre usage (of particular fibres) were considered: (i) zero fibre usage, (ii) low fibre usage (1-39 per cent), (iii) intermediate fibre usage (40-79 per cent, and (iv) high fibre usage (80-99 per cent). When the tendency of yarn producers to alter fibre mixtures was compared with their percentage fibre usage an important factor emerged; those firms that did alter fibre mixes tended to be intermediate to heavy users of wool and low to intermediate users of man-made fibres.

Another interesting feature was the apparent association between fibre substitution and the change in the overall proportions of wool and man-made fibre used in yarn manufacture. Of those 20 respondents who actually varied their fibre mixes, 10 per cent indicated that the proportion of wool used had increased, 50 per cent indicated it had decreased and 35 per cent indicated that it was unchanged. On the other hand, of these 20, 45 per cent indicated that the proportion of man-made fibres used had increased, 10 per cent indicated it had decreased and 40 per cent indicated no change. This is consistent with a general trend of substitution of man-made fibre for wool. The statistical significance of the above relationships was confirmed by chi-square tests of independence,[5] the overall pattern of which is reported in Table A7.1.

In the case of fabric manufacturers, 38 responded to the question[6] on fibre substitution and of these, 16 or 42 per cent indicated that they did alter the proportion of natural and man-made fibre in their manufacture of fabric. As with yarn production, the firm's particular fibre usage (in percentage terms) was compared with actual fibre substitution patterns. The four categories of percentage fibre usage (of particular fibres) were again considered. It appeared once more from the survey, that those firms which did alter fibre mixes were primarily intermediate to high wool users and low to intermediate users of man-made fibres. These possible relationships were tested for statistical significance using the chi-square tests of independence and as Table A7.1 indicates, the results once more proved positive.

---

[4] See Question 6, Appendix to this Chapter.

[5] Chi-square tests of independence were utilised through the University computer to test the statistical significance of the relationships which appeared evident from first observation. Basically, these tests show whether there are any implied relationships between any two sets of variables. Due to the small number of observations encountered in our independence testing, any conclusions reached using chi-square analysis should be regarded as tentative.

[6] See Question 25, Appendix to this Chapter.

Of the 16 respondents who did alter fibre mixes, 25 per cent indicated that they had increased the proportion of wool in fabric, 63 per cent indicated that they had decreased the proportion of wool, while 12 per cent indicated no change. By contrast, 75 per cent of these 16 indicated that the proportion of man-made fibres used in their fabric had increased.

It is evident from our survey that fibre substitution possibilities exist and are, in fact, utilised in the Australian textile industry. In both yarn and fabric production, it appears that those manufacturers who do alter mixes tend to be intermediate to heavy users of wool and low to intermediate users of man-made fibres.

Factors Influencing Fibre Substitution

Past results: Several studies, completed over the past 15 years, have attempted to isolate the reasons behind the changing pattern of fibre consumption. Most of these have concentrated on the use of man-made fibres in the wool textile industry of various countries but some have considered competition between man-made fibres and cotton.

These studies are limited in scope.[7] The United States is the only country for which detailed and reliable end-use consumption figures are available covering all the main textile fibres. However, the International Wool Secretariat provides detailed information on fibre consumption in eight wool end-uses of the main wool textile producing countries. In addition, studies undertaken in other countries (including Australia) provide some additional data on fibre substitution.

These earlier studies of fibre substitution indicate that factors such as technological developments, promotional efforts by synthetic fibre producers, fashion changes, production cost advantages and changes in relative fibre prices influence fibre competition and substitution. Let us consider each of these factors in turn.

Technological development: The range of available synthetic fibres has expanded and the diversity of available characteristics in these fibres has widened as a result of technical progress. Furthermore, the availability of fibres with new characteristics has stimulated technological progress in textile production, much of which has favoured man-made fibres.

Sault [37], suggests that the properties of synthetics are not necessarily characteristic of wool or other natural fibres but have been found desirable by manufacturers and consumers. Easier care, superior wash and wear properties and greater tensile strength of many synthetic fibres have been a product of the

[7] The lack of suitable data has restricted detailed research mainly to the United States and E.E.C. countries. For an explanation of the data problems associated with textile fibre research see Sault [37, pp. 4-9, 68-69].

technological advances by man-made fibre producers. Today some synthetic fibres possess desirable characteristics such as resistance of biological and chemical decomposition of fibre substitution not available in natural fibres.

Earlier studies of fibre substitution, [8], [9], [18], [33], [34], [35], indicate that technological developments combined with complementary alterations in properties of man-made fibres have been a major factor influencing trends in fibre competition and substitution. Minford in a recent study of the United States textile industry found that technological factors accounted for fibre substitution in some instances. He points out that "the displacement of cotton and wool by synthetic yarn in several knitwear end-uses appeared ... to be due ... to the progress of machine-knitting technology" [30, p. 189].

K.C. Clarke [11], formerly Director of the International Wool Secretariat, has emphasised that downstream technological developments in the manufacture of textiles may favour man-made fibres. He elaborates on this point in a recent article and concludes that:

> The dangers to wool ... [are] that the total impact of technological change - methods of fabrication, colouration, garment manufacture etc., - far from being neutral between fibres, may strongly disadvantage wool [11, p. 71].

Promotional efforts by synthetic fibre producers: Despite the technical advances favouring man-made fibres and improvements in and extensions of the properties of man-made fibres (particularly synthetics), the use of synthetics in end-uses traditionally occupied by natural fibres would not have expanded as much as it did without the voluminous and sustained expenditure on promotion undertaken through the 1960s by producers of man-made fibres. Until relatively recently (prior to the increased promotions campaigns such as those associated with Woolmark and Woolblendmark), wool promotion was over-shadowed by the volume of similar expenditures by producers of synthetic fibres.

Fashion changes: Consumer preferences, reinforced by fashion changes, appear to have moved in favour of lighter weight garments. This has favoured strong synthetic fibres. To a considerable extent both fashion changes and alterations in consumer preferences may be a result of the technical improvements and promotion by the man-made fibre producers. One important feature of fashion changes has been the trend towards the use of more casual clothing. This may be a result of significant growth in suburban living, higher living standards and more leisure time. This trend has lead to a marked increased demand for sports-type clothing. Studies by Gruen and Coutts [17], and Jenkins and O'Toole [25], in the early 1960s found that fashion effects were a significant factor in fibre substitution for a number of important end-uses in the United States wool textile industry. Two other studies [8], [9], have also shown that in both the United Kingdom and the E.E.C. markets, fashion changes have been a major factor resulting in the significant penetration of synthetic fibres into established wool end-uses. More recently Clarke [11] has also pointed out that fashion changes have an important bearing on fibre substitution.

Cost advantages in production: As a result of the advent of man-made fibres, significant changes occurred in the costs of converting fibre to yarn and yarn to fabric. Since the cost of converting synthetic fibres to textiles is lower than for natural fibres, this has favoured substitution of the former for the latter. This is suggested in a recent Australian Wool Corporation report which notes that:

> The relative complexity and hence high cost structure of the wool yarn conversion route, particularly for worsted yarns, imposes a significant premium on wool yarn when compared to its competitors [4, p. 48].

This report gives conversion-to-yarn costs of wool, acrylic and polyester fibre. These costs are set out in Fig. 7.3 and broken down by the stages involved in processing fibres to yarn.

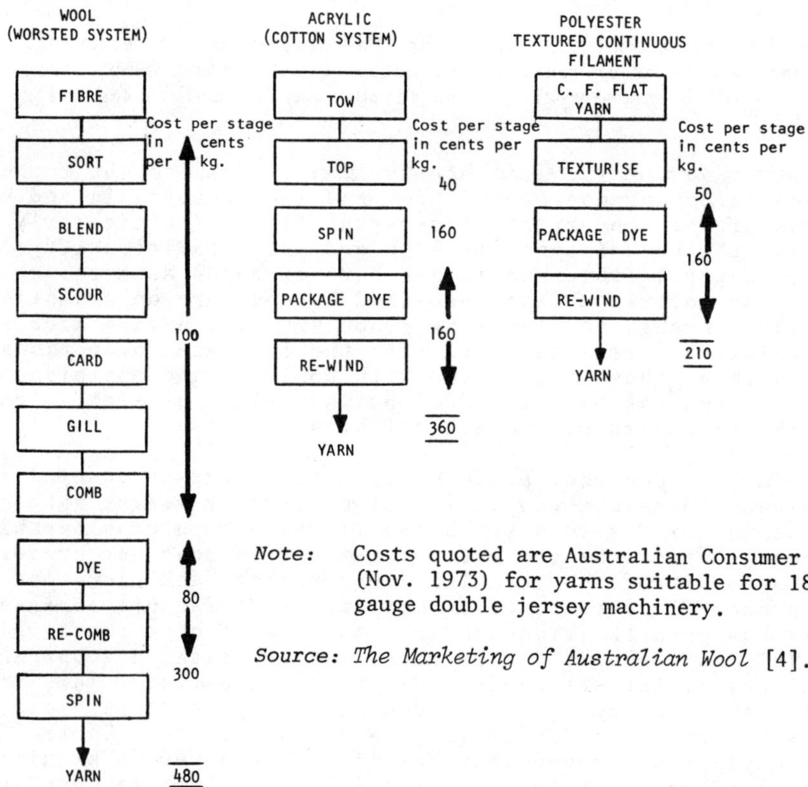

Note: Costs quoted are Australian Consumer basis (Nov. 1973) for yarns suitable for 18 gauge double jersey machinery.

Source: The Marketing of Australian Wool [4].

Fig. 7.3

Stages of Fibre Processing - Raw Material to Yarn

Sault [37], noted the rapid increase in the volume of knitted rather than woven apparel and the sharp rise in the output of tufted rather than woven carpet in the 1960s, and observed that this was adverse to the demand for wool because woven items in both areas traditionally used considerable amounts of wool. Knitting technology favoured synthetics, for instance, synthetic fibres allowed a greater speed of textile operation without breakage of fibres, and this lowered the cost of knitted items considerably.[8]

Synthetic fibres were further advantaged by the fact that some of the less costly processing techniques were developed specifically for production involving only synthetic fibres. Pender found that changes in relative conversion costs in Australia as well as in the U.K. and the U.S.A., "seem to have favoured other fibres, particularly synthetics at the expense of wool" [32, p. 20]. He pointed out that this was most evident in the movement to knitting and tufting at the expense of weaving. The importance of production costs for the substitution of synthetic for natural fibres has also been emphasised by Sturgess [40], [41].

Changes in relative fibre prices: The importance of relative fibre prices as a factor in fibre substitution seems to have increased substantially since the early 1960s. Several of the earlier studies mentioned above concluded that relative fibre price changes were of minor importance in fibre substitution whereas later studies, particularly that of Minford [30], find that changes in relative price are the principal determinants of fibre substitution in most end-uses.

Research conducted in the mid-1960s at the University of Adelaide [33], [34], suggests that although relative fibre prices were not an important factor in fibre substitution in the countries studied[9] (primarily due to excess man-made capacity at the time) relative fibre prices should emerge as an important factor in fibre substitution when full capacity in man-made fibres is approached and "should synthetic prices decline secularly relative to wool, it can be anticipated that this would induce further synthetic substitution" [33, p. 36].

Ward [47], in a study of interfibre competition in the United States concluded that relative fibre prices were not significant in determining substitution between cotton and man-made fibres. In his view non-price aspects such as fibre quality and availability are more important determinants of relative fibre consumption. However, Ward notes that there is evidence of price competition between fibres in the short-term and that cotton, while

---

[8] For a more comprehensive and technically oriented look at conversion costs in high-speed weft knitting see Knapton's article [27].

[9] In their study Polasek and Powell [34] looked at aspects of fibre competition in the U.S.A., U.K., France, West Germany, Italy and Japan. Polasek's study [33], also considered these six countries but concentrated primarily on U.S. experience.

it exhibits negligible over-all price elasticity "is influenced positively by changes in cellulosic and non-cellulosic fibre prices" [47, p. 223].

While expressing a similar view to that of the Adelaide economists, Sault [37], points out that the influence of relative fibre prices is more important after 1962 than prior to this. The increasing importance of relative prices was underlined by the findings of Emmery who in 1967 pointed out that in the U.K. wool end-use market "the price elasticity [of wool] has increased over the study period [1952-1964], because of the increased availability of the synthetics as substitute fibres" [15, p. 46].

Minford has recently completed an important study of fibre substitution. In his study of fibre substitution in the United States textile industry between 1949 to 1968, Minford examines relative fibre prices and the market share of different fibres and finds that:

> ... while technical change and diffusion may have explained changing fibre shares in certain usually specialised end-uses, and contributed to the explanation of others, in most end-uses relative price change has been an important factor [30, p. 175].

He emphasises the importance of relative prices in determining substitution between fibres and his conclusion contrasts with that of Powell, Polasek and Burley [35]. He concludes "that the progress of a fibre within an end-use, once it has made the quantum-jump of penetrating the end-use, depends primarily and directly on price movements" [30, pp. 176-177].

Some of the major factors affecting fibre substitution are covered by Smith and Dardis [39] who recently undertook a study of inter-fibre competition in the United States. They note that:

> The synthetic fibre producers are few in number, operate in a highly technical research orientated environment and have the ability to combine all three variables - price, quality and promotion - in an optimum manner [39, p. 215].

All three variables appear to be important in inter-fibre competition. Hakim [22] arrives at similar conclusions about competition and substitution between man-made fibres and cotton in the United States and observes that:

> The high degree of concentration and vertical integration in the man-made fibre industries around the world have made some indirect benefits possible. These industries benefited from economies of scale, and the subsequent reduction in cost of production of the fibre. Huge investments in research were also made possible which, in turn, made technical improvements in existing fibres and the development of new fibres possible. Research has also led to technical development in the production processes of these fibres ... [and] ... besides research, man-made fibre industries benefited a great deal from sales promotion programmes [22, pp. 121-122].

Factors influencing fibre substitution: survey results: The results of our survey indicate that similar factors to those mentioned above influence the type of fibres chosen by Australian textile manufacturers and their substitution of fibres. The textile manufacturers surveyed were asked to state the circumstances which cause them to change the proportions of man-made and natural fibres used in their production. The main factors mentioned were price factors, changes in consumer preferences, changes in fashion trends, differences in production costs and desirable technical properties. Let us consider each of these factors in turn.

Price factors: Price instability of natural fibres, particularly of wool relative to man-made fibres, was the factor mentioned most frequently by respondents as one influencing their choice and substitution of fibres. In addition, several manufacturers said that due to the lower price of such man-made fibres as nylon (relative to wool) they were using higher proportions of nylon than required to produce favourable strength characteristics in their products. These results accord with some of the findings of previous writers [15], [30], [37].

Changes in consumer preferences: The changing patterns of consumer preferences were emphasised by several textile manufacturers who said that they had increased their use of synthetic fibres to some extent because of changing consumer requirements for such physical properties as weight, wear and tear characteristics. This result accords with results from earlier writers such as Gruen and Coutts [17], and Jenkins and O'Toole [25], and more recently Clarke [11] who contends that changing consumer preferences help to account for fibre substitution, and that these have been influenced by technical advances and promotion undertaken by producers of man-made fibres.

Changes in fashion trends: This factor was also regarded as important by textile producers in the survey. One yarn manufacturer for instance said that fashion trends had an important indirect effect on his choice of fibre. He found it necessary to vary the fibre mixture of his yarns to satisfy changing fashions. He added, however, that this effect was of only secondary importance to such effects as relative fibre prices.

Differences in costs of production: Several textile producers indicated that their production (conversion) costs are lowered by the use of man-made fibres and that this has influenced them to substitute man-made fibres for natural fibres.

Technical properties: Some manufacturers stated that desirable properties can be obtained by substituting man-made fibres for natural fibres in both yarn and fabric production and this had influenced them to increase their use of synthetic fibres. Examples of desirable properties given by respondents included the additional strength and aesthetic qualities gained by either completely substituting man-made fibres for wool or by blending wool and man-made fibres.

TABLE 7.2

Factors Influencing Fibre Substitution as Indicated by Textile Manufacturers Responding to Australian Survey

Factors influencing fibre substitution	No. of firms mentioning this factor	Per cent[a]
1. Price factors:		
(i) Price Instability	11	39.0
(ii) General	8	29.0
2. Changes in Consumer Preferences	10	36.0
3. Changes in Fashion Trends	7	25.0
4. Differences in Costs of Production	4	14.0
5. Technical Properties	8	29.0

Note: (a) Number of firms mentioning a particular factor as a percentage of total number of firms responding to the question on fibre substitution. These percentages do not add to 100 per cent because most respondents gave more than one factor.

Source: First survey.

Table 7.2 summarises the reasons given by Australian textile producers for substituting man-made fibres for natural fibres. It is clear from this Table that most firms (68 per cent) consider price factors to be very important in fibre substitution. These results are similar to those found in overseas studies mentioned earlier and accord with the results of Pender's 1967 study [32] of fibre competition in Australia which found that movement in relative fibre price, changes in production costs (conversion costs) and changes in social habits (for example, fashion, the move to out-door living) seem to favour other fibres, particularly synthetics, at the expense of wool.

7.3 PRICE INSTABILITY AND ITS ROLE IN FIBRE SUBSTITUTION

Price Instability and Fibre Substitution Generally

The last section indicates that relative price changes play an important role in fibre substitution and that such changes have adversely affected the demand for natural fibres. Attention is now focused on the extent to which actual fibre price instability (notably that of wool) influences fibre substitution. Hirsh and Ellis [23] suggest that the possibilities of fibre substitution available to the wool textile industry are now considerable and predict that wool prices will continue to fluctuate widely. Instability of wool prices is likely to be a continuing problem.

Instability of natural fibre prices, particularly wool, creates a considerable degree of uncertainty. This has a major bearing on the decisions of textile manufacturers and, as the Australian Wool Corporation pointed out in its recent report [4], can lead to fibre substitution. This uncertainty, primarily a result of price instability, is borne out by our Australian survey, the results of which are now examined.

Our survey results show that *an association exists between the attitudes of textile manufacturers to price stabilisation and their ability to substitute fibres* in their production. Seventeen (89 per cent) out of the 19 yarn manufacturers who were able to alter their fibre mixes indicated that stabilised fibre prices would be an advantage to them. Chi-square tests of independence indicate that this result is statistically significant.

This result is interesting because it implies that firms with greater flexibility in fibre substitution technology are more readily prepared to pay for stability of wool prices. Furthermore, one would expect that where textile manufacturers have greater ability to alter fibre mixes they would experience greater regret when they make wrong decisions. For instance, if textile manufacturers expect or predict wool-rich mixes to be less profitable than in the past and utilise their machinery to produce textiles containing a low proportion of wool, they will regret such a decision if the returns of wool-rich products are high. Their "regret" is likely to be greater the greater their scope for producing wool-rich products.

The overall effect of price instability on the decisions of textile manufacturers is of particular interest since the Australian Wool Corporation reported in *The Marketing of Australian Wool* [4] that price instability is a significant factor influencing textile producers to substitute man-made fibres for wool. In our survey, we asked Australian textile manufacturers about their attitudes to instability of the prices of fibres.

Yarn Producers and Price Instability: Survey Results

Thirty-two yarn producers responded to our question asking them whether or not stabilised fibre prices would be an advantage to them.[10] Of these, twenty-three (72 per cent) indicated that stabilised fibre prices would be an advantage in the production of yarn. However, on the question whether they would be prepared to pay a premium for this stability, only eight of the 28 respondents (29 per cent) were prepared to pay a premium on wool to guarantee a known stabilised price and the percentage premium per kilogram suggested by these firms ranged from five to ten per cent.

For analytical purposes, yarn producers were classified by their percentage use of a fibre in production. Four categories of percentage use (of particular fibres) were considered; (i) low fibre usage (1-39 per cent), (ii) intermediate fibre usage (40-79 per cent), (iii) high fibre usage (80-99 per cent), and (iv) complete fibre usage (100 per cent).

---

[10] See Question 15, Appendix to this Chapter.

A majority of users of each fibre type (wool, cotton and man-made fibres) and particularly wool users, favoured stabilised fibre prices. Of the 21 producers of yarn using wool, nineteen (90.4 per cent) replied in favour of stabilised prices. Of those nineteen, four were in the low wool usage category, seven in the intermediate wool usage category and eight were in the high usage category. The results for wool are detailed in Table 7.3.

TABLE 7.3

Attitude of Wool Users to Stabilised Fibre Prices - Yarn Producers

Wool as a % of all fibres used	Stabilised fibre prices would be an advantage		Stabilised fibre prices would make no difference		Sub-total No.	Sub-total %
	No.	% of total	No.	% of total		
1 - 39	4	19.0	-	-	4	19.0
40 - 79	7	33.3	1	4.8	8	38.1
80 - 99	8	38.1	1	4.8	9	42.9
100	-	-	-	-	-	-
Sub-total No.	19		2		21	
Sub-total %		90.4		9.6		100.0

Source: First survey.

Similar comparisons were made for yarn manufacturers using various percentages of cotton and man-made fibre in the manufacture of yarn. The main results were: (i) in the case of firms using cotton in various percentages in their production of yarn, 75 per cent regarded stabilised fibre prices as an advantage, (ii) in the case of firms using man-made fibre in various percentages in their production of yarn, 72 per cent regarded stabilised fibre prices as an advantage. Compared to the proportion of yarn producers using wool who considered stabilisation of fibre prices to be advantageous (90.4 per cent), a smaller proportion (72 per cent) of yarn manufacturers using man-made fibres regarded stabilisation of fibre prices as an advantage. Possibly the latter felt that the prices of their fibres were already relatively stable or felt that across-the-board stabilisation of fibre prices would relatively advantage users of natural fibres.

Our survey results indicated an association between the preference of yarn manufacturers for stabilised fibre prices and the actual variation in the proportions of natural and man-made fibres used by them. Of the 20 yarn manufacturers who did vary their fibre mixes, 17 (85 per cent) indicated that stabilised fibre prices would be regarded as an advantage by them.

As noted earlier, this association seems logical because firms which are able to, and actually vary fibre mixtures, firms with the greatest flexibility, are liable to experience the greatest regret when their predictions go astray. The psychological preference of the managers of such firms for price stability is understandable. Furthermore, the costs on average to the firm of price uncertainty, its effect on the level of average profit of the firm is liable to be greater on average for a firm which is more flexible in its production plans [43], [45]. Price stabilisation, by reducing uncertainty, may be more advantageous to yarn producers with greatest flexibility in their fibre mixtures.

A closer look at the above 17 yarn manufacturers reveals that those firms engaging in fibre substitution and also favouring fibre price stabilisation tended to be intermediate to high users of wool and low to intermediate users of man-made fibres. Hence, makers of predominantly wool yarns might be induced to use more wool if its price is stabilised.

The results of chi-square tests shown in Table A7.1 support as statistically significant all of the price instability relationships mentioned above.

In response to the question of whether or not they would be prepared to pay a premium above current prices to buy wool (wool tops or wool inputs) at a known stabilised annual price guaranteed a year in advance,[11] 38 per cent of yarn producers using wool said they would be prepared to pay a premium, 29 per cent said that they did not know whether they would be prepared to pay a premium and 33 per cent said that they would not be prepared to pay a premium. The results are set out in Table 7.4.

TABLE 7.4

Attitudes of Wool Users to a Forward Stabilised Price for Wool-Yarn Producers

Wool as a % of all fibres used	Would be prepared to pay a premium		Would not be prepared to pay a premium		Do not know whether would be prepared to pay a premium		Sub-total No.	Sub-total %
	No.	% of total	No.	% of total	No.	% of total		
1 - 39	2	9.5	-	-	2	9.5	4	19.0
40 - 79	3	14.3	4	19.0	1	4.8	8	38.1
80 - 99	3	14.3	3	14.3	3	14.3	9	42.9
100	-	-	-	-	-	-	-	-
Sub-total No.	8		7		6		21	
Sub-total %		38.1		33.3		28.6		100.0

Source: First survey.

[11] See Question 14, Appendix to this Chapter.

Those indicating their willingness to pay a premium, a stabilised wool price guaranteed a year in advance, suggested that they would be prepared to pay a percentage mark-up on current prices of 5 to 10 per cent per kilogram. Given the spread of responses, it is not certain that yarn producers using wool would be prepared to pay on balance a net premium for guarantees about the future price of wool and maintain their use of wool. Some are, however, prepared to pay a premium and the newly emerging policy of the Australian Wool Corporation [5, p. 2] in arranging forward contracts with individual buyers should be advantageous to this group. It is prudent also to note that responses to this part of our questionnaire could have been influenced by respondents bearing in mind the possible policy applications of the answers.

Fabric Producers and Price Instability: Survey Results

Thirty-seven fabric producers responded to the question asking whether or not stabilised fibre prices would be an advantage to them.[12] Of these, 84 per cent indicated that stabilised fibre prices would be an advantage in the production of fabric.

Classifying fabric manufacturers by the percentage weight of a fibre used in the production of fabric a result similar to that for yarn producers emerges from the survey; a majority of fabric producers using each fibre type, and particularly wool users, regard stabilised fibre prices as an advantage. These results are tabulated in Table 7.5.

TABLE 7.5

Attitudes of Wool Users to Stabilised Fibre Prices - Fabric Producers

Wool as a % of all fibres used	Stabilised fibre prices would be an advantage		Stabilised fibre prices would not be an advantage		Stabilised fibre prices would make no difference		Sub-total No.	Sub-total %
	No.	% of total	No.	% of total	No.	% of total		
1 - 39	11	44.0	-	-	1	4.0	12	48.0
40 - 79	8	32.0	-	-	1	4.0	9	36.0
80 - 99	3	12.0	1	4.0	-	-	4	16.0
100	-	-	-	-	-	-	-	-
Sub-total No.	22		1		2		25	
Sub-total %		88.0		4.0		8.0		100.0

Source: First survey.

[12] See Question 24, Appendix to this Chapter.

Of the 25 fabric manufacturers using wool, 22 said that stabilised fibre prices would be an advantage. Of these 22, eleven were in the low wool usage category, eight were in the intermediate wool usage category, eight were in the intermediate wool usage category, while three were in the high wool usage category.

The attitudes to price stabilisation of fabric manufacturers utilising various percentages of cotton and man-made fibre in their manufacture of fabric were also obtained. In brief, the main results were: (i) in the case of firms using cotton in various percentages in their production of fabric, 93 per cent regarded stabilised fibre prices as an advantage; (ii) in the case of users of man-made fibres, 83 per cent indicated that stabilised fibre prices would be an advantage - a slightly lower percentage than for wool and cotton users.

From the chi-square tests only one statistically significant relationship appeared to exist between fabric producers' preference for stabilised fibre prices and other factors. This was between their preference for stabilised prices and the change in their direct per unit costs of production as full capacity is approached. In particular, 90 per cent of the 31 respondents who said that stabilised fibre prices would be an advantage to them also indicated that they experienced falling direct costs as their production expanded towards full capacity. Of these, 22 were wool users.

Topmakers and Price Instability: Survey Results

Of the four respondent topmakers, two indicated that stabilised fibre prices would not be an advantage to them while the other two indicated that stabilised fibre prices would make no difference. In addition, three out of the four noted that they would not be prepared to pay a premium above current prices to be able to buy wool at a known stabilised annual price guaranteed a year in advance. The fourth topmaker was unsure as to whether he would be prepared to pay a premium or not.

Although this sample of topmakers is small, it seems that the majority of topmakers are not in favour of stabilisation of fibre prices, a result which differs from that for yarn producers and for fibre manufacturers who overwhelmingly considered fibre price stabilisation to be an advantage. Traditionally topmakers appear to have played a role as risk-takers and speculators in the textile industry, a role for which they have naturally received a return. Stabilisation of fibre prices would eliminate or seriously reduce this role of topmakers and their income from speculation. Further attention will be given to this issue in the next Chapter.

Reasons for Favouring Price Stabilisation: Survey Results

Textile producers favouring stabilisation of fibre prices were asked why they expressed such a preference. Most indicated that instability of fibre prices resulted in uncertainty and this adversely affected their decision-making. The reasons given by textile manufacturers for preferring fibre price stabilisation were that such stabilisation would facilitate both planning and control, and possibly lead to more favourable competition with synthetics. Let us consider each of the reasons given by textile manufacturers in favour of stabilisation of fibre prices.

Facilitation of planning: Twenty-three yarn and fabric manufacturers put forward a reason which we classified under this heading. Yarn and fabric manufacturers frequently said that stabilised fibre prices would be an advantage because risk factors associated with unstable and uncertain fibre prices would be minimised. Particular points mentioned were the following: (i) Stabilised fibre prices would facilitate long-run planning. Forward selling would be encouraged and firms would be in a better position to give a price on future orders. Confidence for larger orders would also be encouraged. (ii) Market planning would be more effective. (iii) Medium-term stability would more easily be attained. (iv) There would be more reliable profit estimation. (v) The planning of working capital would be facilitated. (vi) Contract price rises could more easily be avoided. (vii) Stabilised fibre prices (particularly wool) could assist in the establishment of a more stable price structure on a long-term basis particularly in respect to raw material values. (viii) Longer production runs would be possible and this would facilitate cost savings.

Facilitation of control: Eleven firms gave this as a reason. Control can, of course, be related to planning but enough firms listed planning and control as separate reasons to make this distinction here. Firms suggested that stabilised fibre prices would facilitate the control of: (i) raw material purchases, (ii) product, (iii) financial commitments, (iv) imports and local production, (v) costs, and (vi) stock, particularly work-in-progress.

Possibility of more favourable competition with local and imported synthetic yarn and fabric: Three firms gave this as their reason for favouring stabilised prices. Two of these were intermediate to high users of man-made fibres - one in yarn (99 per cent) and the other in fabric (52 per cent) and the third a yarn and fabric producer using predominantly wool (80 per cent). Both the yarn manufacturer and the yarn and fabric manufacturer stressed that keen competition was being encountered from imported fabrics and that stabilised fibre prices, particularly for wool, would enable them to compete more successfully. The yarn manufacturer indicated that in the past he had used wool in his yarn but with fluctuating wool prices he could no longer compete with yarn manufacturers who relied completely on man-made fibres.

Reasons for Not Favouring Price Stabilisation: Survey Results

Two fabric manufacturers and two topmakers indicated that they would regard stabilised fibre prices as a disadvantage. Both fabric manufacturers stressed that they did not favour price stabilisation because of the geographical nature of their production facilities in relation to their supplies of fibre. For example, one of these fabric manufacturers operated in Tasmania and indicated the disadvantage it faced in relation to its mainland competitors. To compete it had to buy wool locally and privately and wool prices fixed at "mainland" rates would be a disadvantage.

The two topmakers who indicated that stabilised fibre prices would be a disadvantage gave similar reasons - profit margins for topmakers were small and they relied on fluctuating markets to take advantage of speculations and so achieve adequate return on funds invested.

Summary of Survey Results

Within the limits of the survey: Most yarn and fibre producers, particularly users of wool, favour stability of fibre prices and a number are prepared to pay a premium for such stability. Textile manufacturers who are able to or do alter their mixtures of different fibres in production more frequently tend to favour price stability. Most manufacturers claim that they support price stability because it facilitates their planning and control of production. On the other hand, a majority of topmakers appears not to favour price stabilisation because it reduces their speculative and risk-bearing role in the marketing process.

7.4 DIRECT COSTS OF TEXTILE PRODUCERS

Theory and Past Empirical Evidence

A number of views have been expressed about the nature of the average costs of production of the typical manufacturing firm. One theory suggests a U-shaped short-run average cost curve and average variable cost (average direct cost, direct per-unit cost or operating cost) curve and stems from the work of Marshall [29]. This is the traditional approach found in microeconomic theory texts such as that by Bain [6]. It assumes that increased output in the short-run results in changes in factor proportions and a decline in marginal product. Rising marginal cost accompanies this declining marginal product and cost curves similar to those in Fig. 7.4 result.

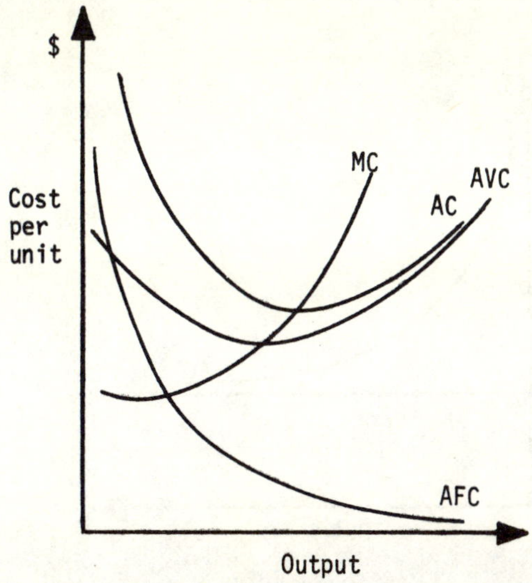

Fig. 7.4

This approach has been criticised as not really showing cost curves as they appear in practice. Bridge and Dodds, for example, note that "empirical evidence is suggestive of the SRAC curve being non-symmetrical but skewed with the minimum level of AC occurring when the plant is around the capacity level" [7, p. 120]. Dean [13] claims that short-run average costs curves of this type seem to be relevant in manufacturing industry.

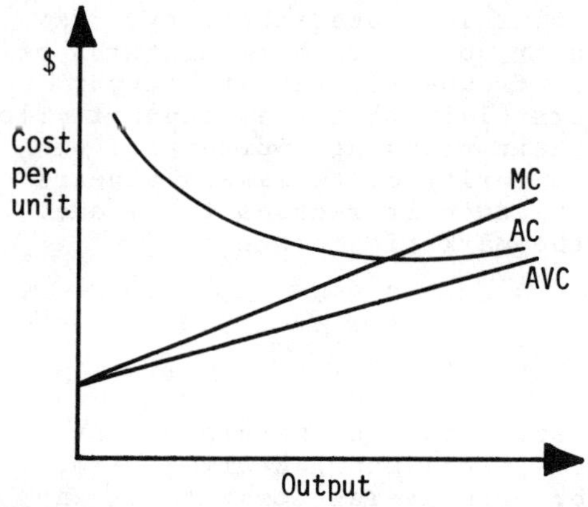

Fig. 7.5

An alternative view, expounded in the work of H.T. Davis [12], claims that cost curves similar to those shown in Fig. 7.5 are typical in manufacturing industry. Here, although average costs are somewhat U-shaped, the average variable cost curve is increasing linearly. Davis, however, does not rely on empirical evidence to support his contention and again empirical evidence seems to suggest that the cost curves of manufacturing firms exhibit different characteristics.

Another view which according to Johnston "is the one suggested as most plausible by the accumulating empirical evidence" [26, p. 13], was originally expressed in the writings of P.W.S. Andrews [1], [2]. It assumes decreasing short-run average costs accompanied by constant marginal and direct per-unit costs similar to those indicated in Fig. 7.6.

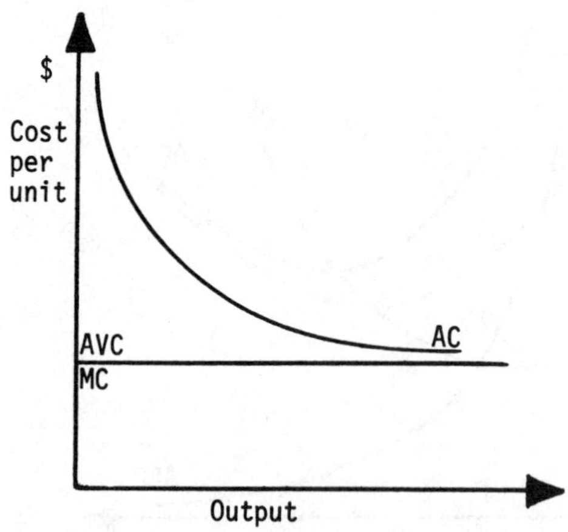

Fig. 7.6

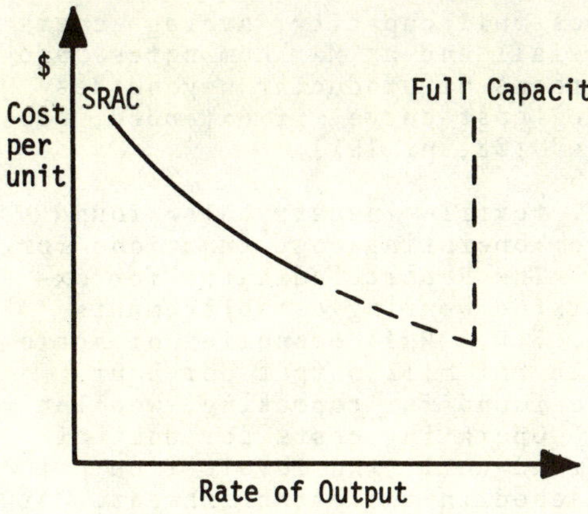

Fig. 7.7

Following his study of a number of manufacturing industries, Andrews [2] concluded that manufacturing firms most commonly experience decreasing short-run average costs of production. A modification of Andrews' model suggests that when full capacity is reached, average costs rise rapidly to form an average cost curve like the one indicated in Fig. 7.7 [7], [28], [38].

The above are general observations about the nature of average costs and average direct costs in manufacturing, but some specific studies have been undertaken of the textile industry. Markham, in his book, *Competition in the Rayon Industry* [28], reports the results of a cost study carried out in the late 1940s by members of the industrial engineering and cost accounting staffs of one of the major multi-plant rayon manufacturers in the United States. Drawing on this study, Markham suggests that the average costs of a typical producer of rayon forms the type of curve depicted in Fig. 7.8.

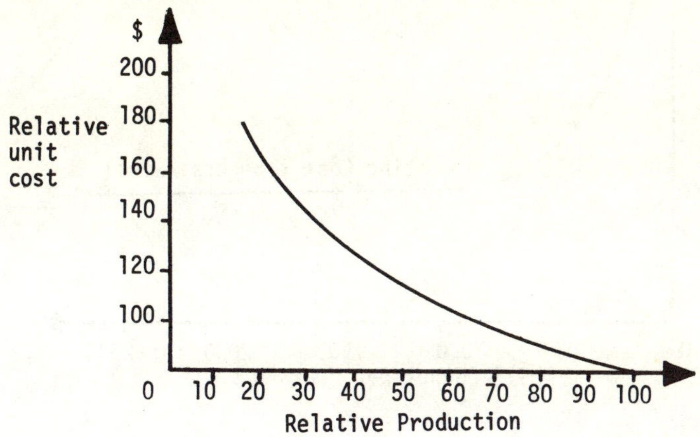

Fig. 7.8

Variations in Average Costs With Percentage of Installed Capacity Utilisation - U.S. Rayon Plants 1949

Source: Based on Chart II, *Competition in the Rayon Industry* [28].

As the firm's production approaches full capacity, average costs and direct per-unit costs tend to fall and as Markham notes "producers do not appear to think in terms of producing beyond designed capacity since the [average] cost curve, if extended, would be almost a vertical straight line" [28, p. 151].

The Atkins Report [3]] on the U.K. textile industry also found downward sloping direct per-unit or operating cost functions for various sections of the industry. The Report indicates for example that operating costs for worsted weaving establishments are similar to those shown in Fig. 7.9. Full economies of scale are achieved in this case at 800 lb. of mill output per hour. Similar operating cost curves were found for topmaking, woollen yarn and worsted yarn production. Operating costs per unit of production display a decreasing trend until the level of output for full economies of scale is reached in the establishment. At outputs in excess of this output, constant operating costs per unit of production prevail. The above results indicate that in order to achieve maximum economic efficiency in the United Kingdom wool textile industry, it is necessary to utilise employed productive capacity fully [14].

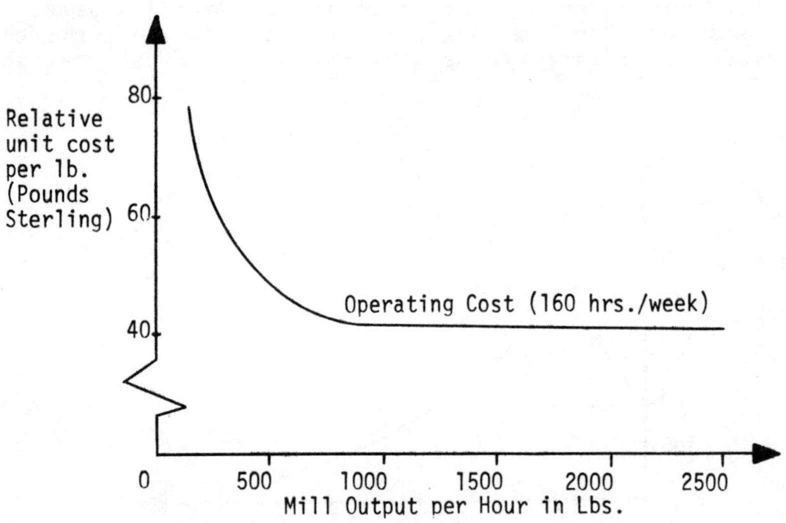

Fig. 7.9

Operating Cost of a U.K. Worsted Weaving Establishment

Source: Based on Fig. 12.5 in *The Strategic Future of the Wool Industry* [31].

Nature of Direct Costs: Survey Results

Textile producers were asked in our survey about the nature of their direct per-unit costs to enable us to compare the Australian position with that reported overseas and to draw possible inferences about the effect of instability of fibre prices on the returns of textile producers.

Firms were asked to indicate the variation of their direct per-unit costs as production moves from surplus capacity towards full capacity.[13] It was found that a majority of responding firms experience falling direct per-unit costs as full capacity is approached; a result which corresponds to the pattern found by Markham [28], in his empirical study, as well as to the results set out in the Atkins Report [31]. The results from our survey concerning the nature of direct costs experienced by yarn manufacturers, fabric producers and topmakers are now summarised.

Yarn producers' direct costs and production capacity: Of the 35 yarn manufacturers responding to this question, 22 (62 per cent) indicated that their direct per-unit costs fall as their production approaches full capacity. Nine (26 per cent) said that their direct costs per-unit altered in an unknown manner while two indicated that these varied in another pattern.[14] From the survey it appears that relatively more man-made fibre users than wool users experience falling direct costs per-unit as full capacity is approached.

Fabric producers' direct costs and production capacity: The majority of fabric producers indicated that their direct per-unit costs of production falls as full capacity is approached. Of the 38 respondent fabric producers 33 (87 per cent) indicated falling per-unit direct costs. One suggested that his overall average variable cost rises, two indicated that their average direct costs remain constant, one said that his average direct costs vary in another pattern and one indicated that his average direct costs varied in an unknown manner.

Proportionately more fabric producers using man-made fibres than those using wool experience falling average direct costs as full capacity is approached. Overall falling average direct cost is relatively more frequent in fabric production than in yarn production.

Topmakers' direct costs and production capacity: Of the three respondent topmakers, one indicated that his direct costs per-unit fall as full capacity is approached and the other two indicated that their average direct costs remain constant.

---

[13] See Questions 16 and 26, Appendix to this Chapter.

[14] These two manufacturers indicated that as their yarn production expanded towards full capacity average direct costs moved in an "inverted saw-tooth" fashion. This, they said, depended on stepped costs but the overall direct cost per-unit tendency was downward. One of these manufacturers elaborated to say that average direct cost "falls to a point where a shift is operating a full capacity and then increases as an additional shift is put into operation".

In the next section these empirical findings about the nature of average costs in the textile industry are used to discuss the possible effects of the stabilisation of fibre prices on the returns of textile producers.

7.5 MORE THEORY: PRICE STABILISATION AND THE RETURNS OF TEXTILE PRODUCERS

Under *conditions of certainty*, textile producers may gain or lose from stabilisation of fibre prices depending upon whether their costs of production increase at an increasing rate or at a decreasing rate with their volume of throughput of fibres [43]. Schemes which stabilise fibre prices are liable to destabilise or increase fluctuations in textile producers' throughput of fibres and this increased variation is liable to lead to greater production costs on average if marginal costs are increasing and to lower costs on average if marginal costs are declining.

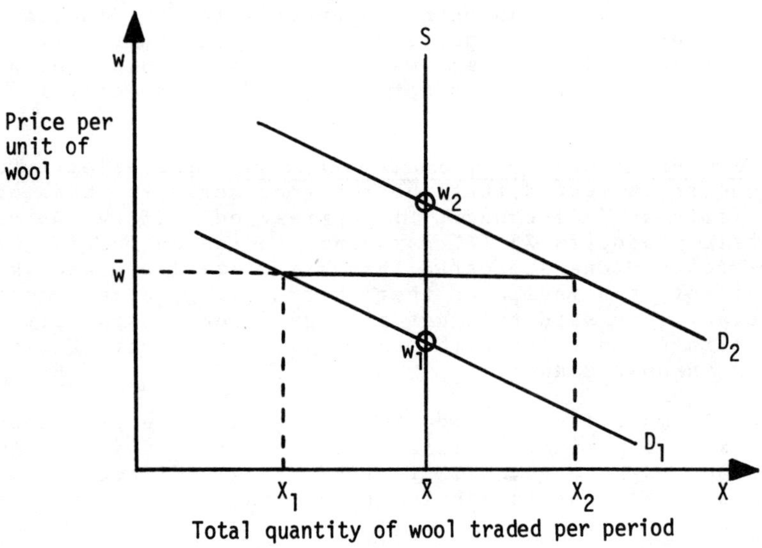

Fig. 7.10

This can be illustrated by the following example. Fig. 7.10 shows the market demand and supply for wool. The supply curve marked S is assumed to be inelastic and the derived demand for wool is supposed to be as indicated by curve D_1 half of the time and as shown by curve D_2 the other half of the time. Thus in the absence of a stabilisation scheme the price of wool fluctuates between w_1 and w_2. But by the use of buffer stocks the price of wool can be stabilised at its average value namely at $\bar{w}$. During periods of low demand a market regulating authority such as the Australian Wool Corporation adds $\bar{X} - X_1$ to its stocks and draws on this quantity during periods of high demand so as to maintain the market price at $\bar{w}$. While this policy stabilises wool prices, it destabilises the quantity of wool released to the trade and handled by textile processors. As a result of stabilisation of wool prices, the quantity of wool handled by the trade is destabilised from $\bar{X}$ per period to an amount varying between X_1 and X_2 per period.

Incidentally, some textile processors have complained about the current wool price stabilisation policies of major wool producing countries because these have destabilised their throughput of wool. A recent article in the *Textile Month* [42] claimed that floor-price schemes for wool are creating difficulties for processors of wool. The article claims that:

> ... if one end of the wool chain is made price safe and moderately comfortable, it means that the other end has to absorb all the buffeting that the market will assuredly administer, *and will have to do it without ever having access to cheap - or even cheaper - wool* [42, p. 26].

Given the relatively high price of wool in 1977 and 1978 in comparison to man-made fibres the demand for wool is less than it would be at a price lower than the floor-price and closer to the market equilibrium one. This means that:

> ... the early processing sector of the U.K. wool-textile industry is very short of orders at present and the only topmakers who are doing any useful weight of business either home-trade or export account are the price-cutters [42, p. 26].

These firms are reputed to be selling wool tops below cost to keep their machinery full utilised and so reduce the size of their losses. "Talk of the advantages of wool price stability in such company makes few friends!" [42, p. 26]. These observed difficulties of processors are consistent with the above theory. A buffer stock scheme destabilises the throughput of processors and increases the variability of their profits. In periods of low demand for textiles, a buffer stock scheme for wool causes the throughput of processors to be lower than it would otherwise be and also results in their profits being lower *at this time* than they otherwise would be.

The effect of greater variability of wool throughput on the incurred costs of textile processors depends upon whether their marginal costs of production are increasing, decreasing or constant.

If textile producers only hold stocks to meet orders[15] (or hold for all practical purposes no stocks) and the costs of these is

---

[15] Evidence given in Chapter 8 indicates that this tends to be so for yarn and fabric manufacturers.

linearly related to output,[16] and if textile producers are equally affected by the aggregate position, their incurred costs rise if their marginal costs are an increasing function of their throughput of wool. On the other hand, the incurred costs of textile processors decline if their marginal costs are a decreasing function of their throughput of wool. If their marginal costs are constant, their total costs on average are unaffected. Fig. 7.11 illustrates these possibilities for a typical textile firm reliant on wool.

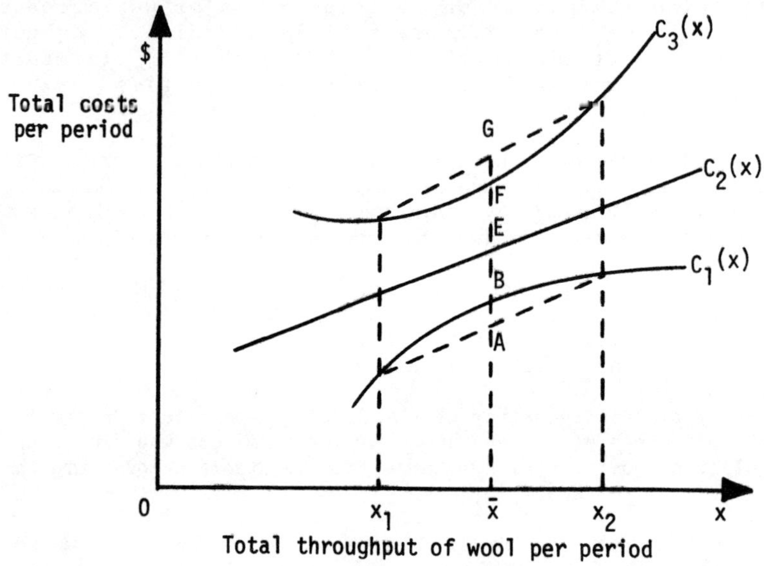

Fig. 7.11

In the case of total cost curve $C_1(x)$ in Fig. 7.11, marginal cost is decreasing at a stable throughput of $\bar{x}$, total costs per period are B. But if throughput is destabilised to become x_1 half of the time and x_2 the other half of the time, total cost per period falls to A on average even though throughput on average remains at $\bar{x}$. In the constant marginal cost case, indicated by total cost function $C_2(x)$, costs on average remain unchanged at E in the same circumstances. In the rising marginal cost case, illustrated by total cost curve $C_3(x)$, costs on average for period rise from F to G if throughput is destabilised under the above circumstances. However, the implications of these variations in costs for the returns of textile producers depends on the way in which their revenue alters. If their revenues per period remain unaltered on average, the implications of price stabilisation for the returns

[16] The marginal costs of stocks in practice is liable to be a declining function of the firms' volume of output. If the pricing relationship discussed below holds, this favours price stabilisation.

of textile manufacturers are clear. But we must consider whether or not revenues are liable to be unaltered.

In order to ascertain the effect of price stabilisation on the revenue of textile producers it is necessary to study their pricing behaviour. Some economists are of the view that cost-plus pricing is common in the textile industry [19], [20], [21], [28], so let us explore the implications of a cost-plus model. To do this, let us take wool as an example and divide the costs of a textile firm relying on wool inputs into two parts, the cost of wool wx and the total cost of processing it J(x), so that

$$C = wx + J(x) \tag{7.1}$$

The per-unit price of the firm's product is based upon the cost of the wool content of each unit of the product plus the other costs assigned to each unit when the firm is operating at *average standard or normal capacity* plus a profit mark-up [7]. Where P is the price of the final product,

$$P = \lambda[w + J(\bar{x})/\bar{x}](1 + \xi) \tag{7.2}$$

when λ is the wool content of the final product, $\bar{x}$ is the firm's standard (average) operating throughput and ξ is the firm's percentage profit mark-up. This mark-up is relatively fixed.

Given the above pricing equation, the price of the firm's final product if its average throughput is maintained and other prices are constant, is only altered by changing wool prices. If the firm's throughput of wool fibres follows a similar pattern to that for the whole industry, its average throughput of wool if left unchanged by the wool price stabilisation scheme illustrated by Fig. 7.10. Consequently, the price which the firm charges for its product *on average* remains at

$$\bar{P} = \lambda[\bar{w} + J(\bar{x})/\bar{x}](1 - \xi) \tag{7.3}$$

and its total revenue *on average* per period remains unaltered at

$$T.R. = P\bar{x} \tag{7.4}$$

because the total revenue curve is linear. Thus with revenue unaltered *on average* the effect of the stabilisation of wool prices on the average returns of the representative textile user of wool depends upon its effect on his level of costs.

Using cost equation 7.1 stabilisation of wool prices raises costs on average, leaves costs unaltered on average, or lowers costs on average depending upon whether J"(x) is positive, zero or negative. If the marginal cost of producing wool is rising, costs on average per period rise; if it is constant they remain unchanged and if marginal cost is falling costs on average per period fall. The argument is analogous to that illustrated by Fig. 7.11.

Hence, under the above conditions, stabilisation of wool prices lowers the returns of textile producers if their marginal costs are rising, leaves returns unchanged if marginal costs are constant and actually raises returns if marginal costs are falling.[17, 18]

Results from our survey mentioned earlier, indicate that the direct per-unit costs of textile producers tend to decline.[19] Although it does not follow from this that the marginal costs of such firms must be declining, the results are consistent with declining marginal costs, and are inconsistent with continually rising marginal costs of production. Thus, under conditions of *certainty* and given the above theory, stabilisation of the price of a fibre such as wool is liable to raise the average returns of textile producers. Although this returns-effect may be small, it is likely to be positive.

Furthermore, in so far as price stabilisation reduces uncertainty, errors in decision-making and the cost of decision-taking by textile firms, this will also add to their returns. Indeed, the most common reason given by textile producers in our survey for favouring stabilisation of future prices was that this would reduce their uncertainty and facilitate their planning, decision-making and control.

Under the above conditions a fibre with a stable price, such as a man-made fibre, is likely to be at a competitive advantage compared to a fibre such as wool which is unstable in price. But the extent of the advantage may depend upon whether the textile processor is principally engaged in production or in speculation. Stabilisation will reduce the income of a successful speculator even if textile processors gain overall as a result of stabilisation. This aspect is explored in the next Chapter.

REFERENCES

[1] ANDREWS, P.W.S., *Manufacturing Business*, Macmillan, New York, 1949.

[2] ANDREWS, P.W.S., "Industrial Analysis in Economics", Ch. 4 in T. Wilson and P.W.S. Andrews (eds), *Oxford Studies in the Price Mechanism*, Oxford, 1951.

[3] Australian Bureau of Statistics, *Monthly Review of Statistics*, Australian Government Publishing Service, Canberra, Sept. 1976 and Dec. 1977.

---

17 Stabilisation of fibre prices has the adverse side-effect of increasing the variability of the returns of textile manufacturers.

18 If $C(z)$ represents a textile manufacturer's total cost of producing his product z, and if the product contains wool in a fixed proportion, $J''(x)$ is, by the function-of-a-function rule for differentiation, of the same sign as $C''(z)$.

19 In our survey [46] we avoided asking textile manufacturers about their changes in marginal costs on the grounds that the concept would be unfamiliar to many and that many firms would not be able to estimate these. It was also necessary not to make our questionnaire too demanding in order to maintain a reasonable response rate.

[4] Australian Wool Corporation, *The Marketing of Australian Wool*, Melbourne, 1973.

[5] Australian Wool Corporation, *Wool News Digest*, June, 2, 1976.

[6] BAIN, J.S., *Price Theory*, Holt, Rinehart and Winston, New York, 1952.

[7] BRIDGE, J., and DODDS, J.C., *Managerial Decision Making*, Croom Helm, London, 1975.

[8] Bureau of Agricultural Economics, *Wool Consumption Trends in Western Europe and the United States*, Wool Economic Research Report No. 3, Canberra, 1961.

[9] Bureau of Agricultural Economics, *Wool in the E.E.C.*, Wool Economic Research Report No. 6, Canberra, 1964.

[10] Bureau of Agricultural Economics, *The Fibre Review 1974-75*, Canberra, 1975.

[11] CLARKE, K.C., "The Future for Wool", *Long Range Planning*, 8, 70-74, 1975.

[12] DAVIS, H.T., *The Theory of Econometrics*, Principia Press, Bloomington, 1941.

[13] DEAN, J., *Managerial Economics*, Prentice-Hall, New Jersey, 1951.

[14] DURBIN, S., *Short Term Price Formation in the Wool Market*, University of New England, N.S.W., 1975.

[15] EMMERY, M., *The Price Elasticity of Demand for Wool in the United Kingdom*, Wool Economic Research Report No. 11, Bureau of Agricultural Economics, Canberra, 1967.

[16] F.A.O. Commodities and Trades Division, *Impact of the Energy Crisis on the Competitive Position of Agricultural Raw Materials*, Rome, 1974.

[17] GRUEN, F.H., and COUTTS, A.M., "An Analysis of Changes in U.S. Wool Consumption", *The Australian Journal of Agricultural Economics*, 5, 93-112, 1961.

[18] GRUEN, F.H., and others, "Changes in Supply of Agricultural Products", in D.B. Williams (ed.), *Agriculture in the Australian Economy*, Sydney University Press, Sydney, 1967.

[19] HAGUE, D.C., "Economic Theory and Business Behaviour", *Review of Economic Studies*, 16, 144-57, 1949.

[20] HAGUE, D.C., *The Economics of Man-Made Fibres*, Duckworth, London, 1957.

[21] HAGUE, D.C., *Pricing in Business*, George Allen and Unwin, London, 1971.

[22] HAKIM, O.A.R., "The Effects of the United States Cotton Policy on the World Market for Extra-Long Staple Cotton", unpublished Ph.D. thesis, University of Arizona, Tucson, 1972.

[23] HIRSH, B.W., and ELLIS, P., *An Introduction to Textile Economics*, The Textile Trade Press, Manchester, 1974.

[24] International Wool Secretariat, *Statistical Summaries*, London, 1971.

[25] JENKINS, E.L., and O'TOOLE, R.J., "Wool in Clothing: U.S.A. and Japan", *Quarterly Review of Agricultural Economics*, 16, 20-28, 1963.

[26] JOHNSTON, J., *Statistical Cost Analysis*, McGraw-Hill, New York, 1960.

[27] KNAPTON, J.J.F., "The Economic Benefits of High-Speed Weft Knitting", *Textile Institute and Industry*, April, 100-102, 1975.

[28] MARKHAM, J.W., *Competition in the Rayon Industry*, Harvard University Press, Cambridge, 1952.

[29] MARSHALL, A., *Principles of Economics*, Macmillan, New York, 1930.

[30] MINFORD, P., "Textile Fibre Substitution and Relative Prices", *The Australian Journal of Agricultural Economics*, 19, 175-196, 1975.

[31] National Economic Development Office (N.E.D.O.), *The Strategic Future of the Wool Textile Industry*, HMSO, 1969.

[32] PENDER, R.W., "Trends in Fibre Competition in Australia With Particular Reference to Wool", *Quarterly Review of Agricultural Economics*, 20, 57-71, 1967.

[33] POLASEK, M., "Synthetic Fibres and Australia's Economic Future", *The Economic Record*, 41, 23-37, 1965.

[34] POLASEK, M., and POWELL, A., "Wool and Synthetics: A Statistical Analysis of Fibre Substitution in the U.S.", *The Australian Journal of Agricultural Economics*, 8, 49-64, 1964.

[35] POWELL, A., POLASEK, M., and BURLEY, H.T., "Synthetic Fibres in the Wool Textile Industry: A Study of the Role of Price in Technological Adjustment", *The Australian Journal of Agricultural Economics*, 7, 107-120, 1965.

[36] ROBERTS, R.C., *A Review of Developments in the Man-Made Fibres 1960 to 1971*, Wool Economic Research Report No. 24, Bureau of Agricultural Economics, Canberra, 1973.

[37] SAULT, J.L., *End Uses of Fibres in the U.S.A.*, Wool Economic Research Report No. 18, Bureau of Agricultural Economics, Canberra, 1970.

[38] SHERMAN, R., *The Economics of Industry*, Little, Brown, Boston, 1974.

[39] SMITH, R., and DARDIS, R., "Inter-Fibre Competition and the Future of the United States Cotton Industry", *The American Journal of Agricultural Economics*, 54, 209-215, 1972.

[40] STURGESS, I.M., "The Wool Board's Second Report on Marketing: A Review Article", *The Australian Journal of Agricultural Economics*, 12, 16-34, 1968.

[41] STURGESS, I.M., "Marketing Policies for Rural Industry" in D. Douglas (ed.), *National Rural Policy*, University of Sydney Extensions Board, Sydney, 1971.

[42] Textile Business Press Ltd., *Textile Month*, August 1977.

[43] TISDELL, C.A., *The Theory of Price Uncertainty, Production and Profit*, Princeton University Press, Princeton, 1968.

[44] TISDELL, C.A., "Implications of Learning for Economic Planning", *Economics of Planning*, 10, 177-192, 1970.

[45] TISDELL, C.A., "Some Circumstances in which Price Stabilisation by the Wool Commission Reduces Incomes", *The Australian Journal of Agricultural Economics*, 16, 94-101, 1972.

[46] TISDELL, C.A., KEATING, G.R., and McDONALD, P.W., *A Survey of Australian Textile Manufacturers: A Short Report of Results for Participating Yarn and Fabric Manufacturers and Topmakers,* The University of Newcastle, N.S.W., 1975.

[47] WARD, L.E., "Interfibre Competition with Emphasis on Cotton", unpublished Ph.D. thesis, University of California, Davis, 1968.

APPENDIX TO CHAPTER 7

TABLE A7.1

Implied Relationships[(a)] from the Questionnaire of Australian Yarn and Fabric Manufacturers, Obtained Using Chi-Square Tests of Independence[(b)]

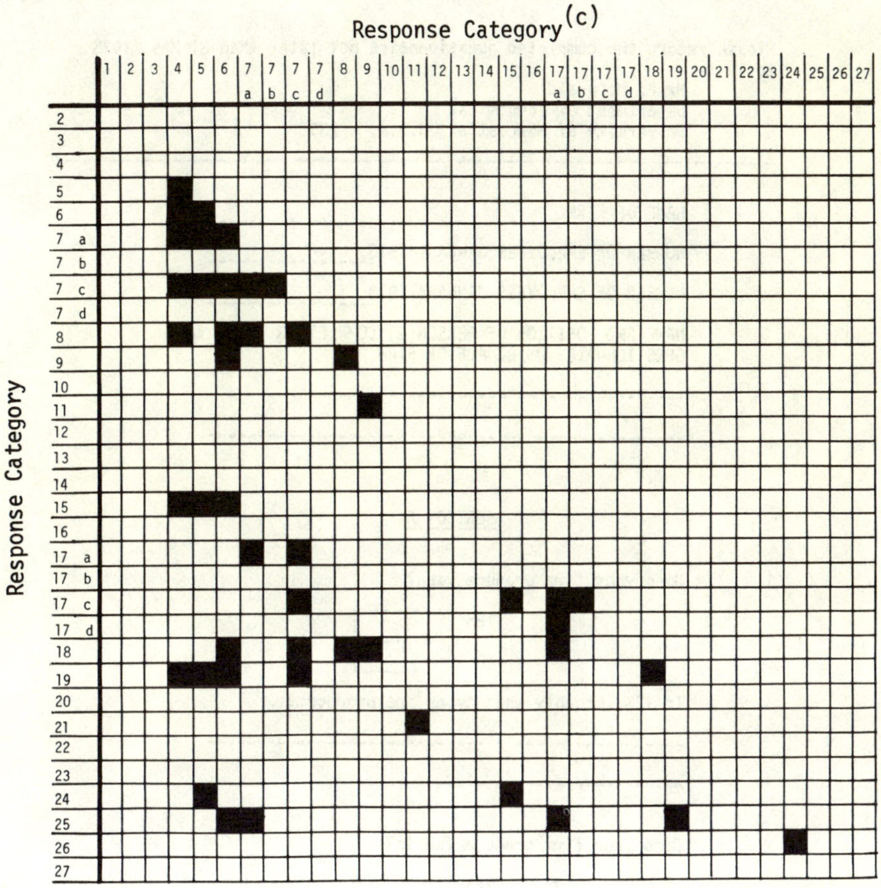

Notes: (a) Not all these implied relationships were considered relevant because the response to some questions was too low to justify dependence in the Chi-Square sense.

(b) Blocked square indicates an implied relationship between the corresponding response categories.

(c) Response category numbers refer to the question numbers in the Questionnaire.

Source: Data from first survey.

QUESTIONNAIRE FOR AUSTRALIAN YARN AND FABRIC MANUFACTURERS

Please return the completed questionnaire not later than 31 May, 1975 to:

Mr P. McDonald
Department of Economics,
University of Newcastle N.S.W. 2308

NAME OF FIRM___________________________________

NUMBER OF EMPLOYEES JANUARY 1975_______________

NUMBER OF EMPLOYEES JANUARY 1973_______________

NAME AND POSITION OF PERSON(S) COMPLETING THIS QUESTIONNAIRE ON BEHALF OF FIRM

____

____

SECTION A

1. Does your firm produce yarn?

YES ☐
NO ☐

If YES, broadly what types are produced?

____

____

2. Does your firm produce fabric?

YES ☐
NO ☐

If YES, broadly what types are produced?

____

____

3. What products besides yarn or fabric does your firm produce?

____

____

SECTION B

(Section B is intended for firms producing yarn. If your firm produces fabric but not yarn, please turn to Section C.)

4. Does your machinery enable you to use wool in your manufacture of yarn?

YES ☐

NO ☐

If YES, and you are not using wool in your yarn, why are you not using it?

----

5. Does your machinery enable you to alter the proportion of man-made fibres and natural fibres contained in your yarn?

YES ☐

NO ☐

6. Do you in fact vary the proportion of man-made and natural fibres which are generally contained in your yarn?

YES ☐

NO ☐

If YES, what circumstances cause you to change your proportions of man-made and natural fibres?

----

7. State the approximate percentage by weight of various types of fibres contained in the firm's output of yarn.

	%
Wool	------
Cotton	------
Man-made Fibres	------
Other Fibres	------

(NOTE: Blends in tops should be allocated according to basic constituent fibres.)

8. The overall <u>proportion</u> of wool input contained in the firm's production of yarn has

increased	☐
decreased	☐
remained unaltered	☐

over the last few years.

Why? --

----

9. The overall <u>proportion</u> of man-made fibre inputs contained in the firm's production of yarn has

increased	☐
decreased	☐
remained unaltered	☐

over the last few years.

Why? --

----

10. If you have substituted man-made fibres for wool, which type of fibre have you mainly substituted?

----

11. The overall <u>proportion</u> of cotton fibre inputs contained in the firm's production of yarn has

increased	☐
decreased	☐
remained unaltered	☐

over the last few years.

Why? --

----

12. **What are the main factors which determine your choice of a fibre or mix of fibres for your yarn?**

----

13. State some circumstances, if any, under which you would be prepared to use a higher proportion of wool in your overall production of yarn.

----

14. Would you be prepared to pay a premium above current prices to be able to buy wool (wool tops or wool inputs) at a known stabilised annual price guaranteed a year in advance?

YES	☐
NO	☐
DO NOT KNOW	☐

If YES, what percentage increase per kg. would you be prepared to pay?

A _______ per cent increase.

15. Would stabilised fibre prices be an advantage to you?

YES	☐
NO	☐
MAKE NO DIFFERENCE	☐

Why? ____

----

16. As the firm's production of yarn expands from a surplus capacity situation towards full capacity, its direct cost (operating or prime cost) per unit of production

(a)	rises	☐
(b)	falls	☐
(c)	remains constant	☐
(d)	varies in another pattern	☐
(e)	alters in a manner not known to us	☐

If (d) please elaborate.

SECTION C

(To be completed by firms producing fabric.)

17. State the approximate percentage by weight of various types of fibres contained in the firm's output of fabric

	%
Wool	-----
Cotton	-----
Man-made Fibres	-----
Other Fibres	-----

(NOTE: Blended yarns should be allocated according to basic constituent fibres.)

18. The overall proportion of wool input contained in the firm's production of fabric has

increased	☐
decreased	☐
remained unaltered	☐

over the last few years.

Why? ---

----

19. The overall proportion of man-made fibre inputs contained in the firm's production of fabric has

increased	☐
decreased	☐
remained unaltered	☐

over the last few years.

Why? ____

____

____

20. If you have substituted man-made fibres for wool, which type of fibre have you mainly substituted?

____

____

21. The overall proportion of cotton fibre inputs contained in the firm's production of fabric has

increased	☐
decreased	☐
remained unaltered	☐

over the last few years.

Why? ____

____

____

22. What are the main factors which determine your choice of yarn?

____

____

____

____

____

23. State some circumstances, if any, under which you would be prepared to use a higher proportion of wool in your overall production of fabric.

----

24. Would stabilised fibre prices be an advantage to you?

YES	☐
NO	☐
MAKE NO DIFFERENCE	☐

Why? --

----

25. Do you vary the proportion of man-made and natural fibres contained in your overall output of fabric?

YES	☐
NO	☐

If YES, what circumstances cause you to change your proportions of man-made and natural fibres?

----

26. As the firm's production of fabric expands from a surplus capacity situation towards full capacity, its direct cost (operating or prime cost) per unit of production

(a)	rises	☐
(b)	falls	☐
(c)	remains constant	☐
(d)	varies in another pattern	☐
(e)	alters in a manner not known to us	☐

If (d) please elaborate.

QUESTIONNAIRE FOR AUSTRALIAN TOPMAKERS

Please return the completed questionnaire not later than 31 May, 1975 to:

Mr P. McDonald
Department of Economics,
University of Newcastle N.S.W. 2308

NAME OF FIRM ____________________

NUMBER OF EMPLOYEES JANUARY 1975 __________

NUMBER OF EMPLOYEES JANUARY 1973 __________

NAME AND POSITION OF PERSON COMPLETING THIS QUESTIONNAIRE ON BEHALF OF FIRM

____

____

1. Does your machinery enable you to alter the proportion of man-made fibres and natural fibres contained in your tops?

   YES ☐
   NO ☐

2. Do you in fact vary the proportion of man-made and natural fibres which are generally contained in your tops?

   YES ☐
   NO ☐

   If YES, what circumstances cause you to change your proportions of man-made and natural fibres?

   ____

   ____

   ____

   ____

3. State the approximate percentage by weight of various types of fibres contained in the firm's output of tops.

	%
Wool	-----
Cotton	-----
Man-made Fibres	-----
Other Fibres	-----

(NOTE: Blends should be allocated according to basic constituent fibres.)

4. The overall <u>proportion</u> of wool input contained in the firm's production of tops has

- increased ☐
- decreased ☐
- remained unaltered ☐

over the last few years.

Why? --

----

5. The overall <u>proportion</u> of man-made fibre inputs contained in the firm's production of tops has

- increased ☐
- decreased ☐
- remained unaltered ☐

over the last few years.

Why? --

----

6. If you have substituted man-made fibres for wool, which type of fibre have you mainly substituted?

----

7. What are the main factors which determine your choice of a fibre or mix of fibres for your tops?

----

8. State some circumstances, if any, under which you would be prepared to use a higher proportion of wool in your overall production of tops.

----

9. Would you be prepared to pay a premium above current prices to be able to buy wool at a known stabilised annual price guaranteed a year in advance?

YES	☐
NO	☐
DO NOT KNOW	☐

If YES, what percentage increase per kg. would you be prepared to pay?

A ________ per cent increase.

10. Would stabilised fibre prices be an advantage to you?

YES	☐
NO	☐
MAKE NO DIFFERENCE	☐

Why? --

----

11. As the firm's production of tops expands from a surplus capacity situation towards full capacity, its direct cost (operating or prime cost) per unit of production

(a) rises ☐
(b) falls ☐
(c) remains constant ☐
(d) varies in another pattern ☐
(e) alters in a manner not known to us ☐

If (d) please elaborate.

THIS SECTION IS FOR ANY GENERAL COMMENTS:

CHAPTER 8

Expectations, Inventories and Use of Futures Markets by Textile Manufacturers

8.1 INTRODUCTION

While previous Chapters have dealt with several aspects of the price instability of natural fibres, especially instability of wool prices, the role of inventories and futures markets in helping textile manufacturers adjust to fluctuations in fibre prices has not been examined. These issues are taken up in this Chapter which, amongst other things, reports results obtained from a supplementary survey [38], of Australian textile manufacturers.

Manufacturers hold stocks of raw wool for various reasons. For instance price instability by affecting price expectations can influence the level of stocks held by textile manufacturers. Our supplementary survey attempted to determine whether textile manufacturers' holdings of wool stocks are influenced by their expectations of wool price movements and by their anticipated sales. This was considered important in itself and also because some Australian textile producers had indicated in our earlier survey [37],[1] that wool price instability interfered with their holding and control of wool stocks (particularly stocks associated with work-in-progress).

In order to reduce the risk associated with wool price variability, manufacturers can also utilise futures markets. A short description of the functions and use of futures markets (wool futures markets in particular) is given in this Chapter.[2] These include:

---

[1] Results from our first survey which deals with other aspects of wool price instability, are reported in Chapter 7.

[2] There have been some extensive changes in the theory of commodity futures markets since the early theoretical work of writers such as Hawtrey [17], Keynes [27], Hicks [18] and Kaldor [26]. At the forefront of those developments have been the work of Working [44], [46] who widened the concept of hedging to encompass more realistic trading behaviour, and other writers such as Brennan [4], Stein [34], and Goss [10], who have made important

(i) the extent to which textile producers use the Sydney Futures Exchange; (ii) the extent to which textile producers believe that the wool futures market enables them to avoid the risks associated with fluctuations in wool prices; and (iii) their opinions of whether the wool futures market plays a useful stabilising role in the marketing of wool.

The supplementary survey also followed up the suggestion that textile manufacturers would prefer a stable market situation [1] and further information was obtained on the attitudes of textile manufacturers to a guaranteed price for wool.

The supplementary survey was undertaken in mid-1976. Sixty-four Australian textile manufacturers were sent a copy of the *Supplementary Questionnaire*.[3] Forty-one returns were received yielding a response rate of 64 per cent. Table 8.1 gives the distribution of responses.

TABLE 8.1

Number of Respondents by Manufacturing Process

Manufacturing process	Total
Yarn and Fabric	10
Yarn	10
Fabric	18
Topmakers	3
Total firms	41

Source: Supplementary survey.

8.2 STOCKS, EXPECTATIONS AND PRICE INSTABILITY

The Holding of Wool Stocks and Price Expectations

Stocks of greasy wool are held by textile manufacturers for a number of reasons: (i) Stocks allow the continuous use of machinery at a capacity necessary to satisfy all the requirements of

contributions to theories of market and individual equilibrium in futures markets. In recent times the short-run speculative price behaviour of futures markets has increasingly come more into focus and in this regard economists such as Samuelson [32], have covered important groundwork. It is not the intention of this Chapter to give any detail on these and other theoretical developments and the reader interested in expanding his knowledge in this regard can find adequate coverage elsewhere [10], [12], [15], [30].

3 A copy of the *Supplementary Questionnaire* is included in the Appendix to this Chapter.

the firm's customers as orders are placed [40]. (ii) Stocks may be held in anticipation of prompt requirements for tops. (iii) Stocks are held also to allow for hedging the sale of tops. When particular orders are received and the topmaker does not want to, or cannot, cover the order immediately, he can hedge his wool by simultaneously buying wool. This wool is turned over frequently and the procedure minimises the cost of the hedge while protecting the initial price margin between tops and wool [8]. (iv) Stocks are held also for speculation on price. Price speculation is intended to serve the dual purpose of profiting the topmaker by putting him in a more profitable position and providing a service to spinners in that cheaper tops are available.

The costs of stock holding have been listed by Carland [6], as the costs of storage and handling as well as the losses due to the tying up of funds in stock which would be invested more profitably elsewhere. On the other hand, he notes that another benefit (in addition to the four mentioned previously) is that by building up stocks, textile firms can gain discounts from bulk buying.

In addition to the factors which directly affect the quantity of wool purchased by textile manufacturers (for example, the price and availability of substitute fibres, research and innovation in fibre and machine technology) another important element is the changing level of wool stocks [1, p. 65].

Stock adjustments by textile manufacturers may be caused by short-term or long-term factors. Increases in wool stocks, for example, may be attributed to three principal factors. The first is the expectation by textile manufacturers that current stocks are inadequate to meet anticipated increases in demand and hence sales. The second factor is that a precautionary upward adjustment of stocks may occur in the belief that forthcoming production is likely to be too low or consumption too high (at current prices) or alternatively in anticipation of some disruption to the supply of wool. A third factor which may contribute to an increase in textile manufacturers' wool stocks is the expectation that future prices will rise so enabling the wool to be eventually resold at a profit as either greasy or semi-processed wool. The Australian Wool Corporation reported that the combined effect of these factors:

> ... is that at a time of rising demand, stock accumulation causes prices to increase more rapidly than future wool requirements would dictate. The extent of the price increase however, tends to be truncated by sales or consumptions out of stocks at a time when expectations of future price rises diminish. An inverse pattern is inclined to accelerate a price fall until expectations are once more reversed [1, p. 65].

Survey results: Using our supplementary survey results, we related the holdings of wool stocks by textile producers to their expectations about wool prices. Each manufacturer was asked whether his holding of wool stocks (in various forms) is influenced by his expectations of wool price movements. Twenty-eight manufacturers responded to this question; 17 (61 per cent) said no and 11 (39 per cent) said yes. Of the 17 producers saying no, seven were yarn and fabric manufacturers, four were yarn manufacturers and six were fabric manufacturers. Of the 11 saying yes, one was a yarn and fabric manufacturer, four were yarn manufacturers, three were fabric manufacturers and three were topmakers. The stocks of all topmakers were influenced by wool price expectations but the stocks of most of the other processors were not.

Reasons given by producers claiming that their wool stocks are influenced by their price expectations: These textile producers put forward five reasons: (i) Three manufacturers said that although stock purchasing is primarily dictated by market demand, stock purchasing is altered slightly on expectations of wool price movements. One yarn manufacturer added that this was dependent upon the price of wool at the time of costing an article to be manufactured in the near future. Two other yarn manufacturers indicated that if there appeared to be a possibility of a rise in wool prices they may buy small quantities of wool ahead of demand to reduce the impact of the price rise. (ii) More wool stocks are held when prices appeared to be on the uptrend. One yarn and fabric producer gave this as his reason for varying wool stocks. (iii) Fewer wool stocks are held when the price of wool is too high to give the processor an adequate return on his manufactured product. The yarn manufacturer expounding this view added that if the above circumstances prevailed and stability of wool prices appeared to exist, he would be inclined to hold higher wool stocks. (iv) When high wool prices are expected, forward requirements for wool to fulfil contracts are purchased at an earlier date than otherwise - two topmakers said this and added that if lower wool prices are expected they purchase wool only as and when required for production. Durbin [8], points out that this tendency exists elsewhere. He found that speculation on price was a reason put by European topmakers for holding greasy wool stocks. (v) Another topmaker said that his wool holding is influenced to a limited extent by his expectations of wool price movement - his action depends in part on his opportunity to hedge on futures.

Reasons given by producers claiming that their wool stocks are not influenced by their price expectations: The most frequent reason given was that stocks are primarily determined by future orders. As such the manufacturers' stocks are not influenced by price expectations about wool. Five manufacturers gave this as their answer, four being yarn and fabric manufacturers (all use wool in high proportions) and one was a yarn manufacturer. Other reasons given by manufacturers were: (i) Insufficient funds prevent stockpiling. Three manufacturers indicated that the present state of the industry was such that their liquidity dictated the barest stock levels be maintained. Durbin [8], also found that limitations on credit restrict stockholding. (ii) Two producers, one yarn manufacturer and one fabric manufacturer, said that their business operations are too small to justify stockpiling of wool. (iii) Costing policy precludes the possibility of stocks being influenced by price expectations. One fabric manufacturer costs on a financial year basis at agreed prices for its end products. Its stock (and forward orders) is for 12 months supply of yarn at the best price it can obtain at the time agreement is entered into. (iv) One yarn manufacturer said that its policy does not provide for speculation about future price movements.

The Holding of Wool Stocks and Anticipated Sales

The relationship between wool stocks of textile producers and their anticipated sales was investigated also in our supplementary survey. In his *Model of the Japanese Wool Textile Industry* Carland [6], postulated that textile manufacturers find it important to maintain sufficient stocks of raw wool to ensure continuous production and to prevent loss of sales in the face of demand fluctuations. He regards his hypothesis, "that stocks do influence demand decisions and dampen demand impulses in relation to raw wool demand" [6, p. 13], as being verified because his empirical result suggests that:

> ... as stocks of raw wool fall producers increase purchases to meet production needs. Conversely as stocks build up production requirements can be more adequately met from stocks ... [6, p. 13].

Survey results: Textile manufacturers were asked to state whether their current wool stocks (April/May 1976) were at normal levels, higher than normal or below normal levels in relation to anticipated sales. There were 28 respondents and of these 19 (68 per cent) said that current stocks were normal, five that they were lower than normal and four that their current stocks in relation to anticipated sales were higher than normal. The distribution of responses in relation to type of manufacturer is outlined in Table 8.2. As will become apparent later, it seems that wool stock holdings of most textile producers are more closely related to their anticipated level of wool sales rather than their price expectations about wool.

TABLE 8.2

Holdings of Wool Stocks by Textile Producers
Classified by Their Manufacturing Process

Current wool stock situation	Yarn & fabric	Yarn	Fabric	Tops	Total
Normal	5	5	6	3	19
Higher than normal	3	1	-	-	4
Lower than normal	1	2	2	-	5

Source: Supplementary survey.

Expectations of Short-Term Wool Price Movements

In our supplementary survey, textile manufacturers were asked to state whether they expected wool prices in the forthcoming year (mid-1976 to mid-1977) to rise, fall or remain unchanged. There were 24 respondents to this question and of these, 16 (67 per cent) said they expected wool prices to rise in the coming year, three said fall and five said remain unchanged. On the whole an upward movement was expected. The distribution of responses in relation to type of manufacturer is outlined in Table 8.3.

TABLE 8.3

Expectations of Short-Term Wool Price Movements by Different Types of Textile Producers

Expected price movement	Yarn & fabric	Yarn	Fabric	Topmaker	Total
Rise	5	5	5	1	16
Fall	-	1	1	1	3
Remain unchanged	2	1	1	1	5

Source: Supplementary survey.

It is interesting to note that 11 of the 16 manufacturers who indicated that they expected wool prices to rise in the coming year also indicated that their wool stock holdings were not influenced by expectations of wool price movements, and 12 of these 16 indicated that current wool stocks, in relation to anticipated sales, were normal. So although the majority expected wool price rises this, apparently, had little influence on their stock holdings which in relation to expected sales were currently at normal levels.

Summarising the results of the supplementary survey in regard to wool inventories, it is clear that most textile manufacturers' holdings of wool stocks are not influenced by their expectations of wool price movements. On the other hand, topmakers said that their stocks are influenced by such considerations. Apparently a manufacturer's wool stocks are less likely to be influenced by his expectations of wool price movements the further he is situated along the wool production chain.

At the time of the supplementary survey most manufacturers were holding "normal" wool stocks in relation to anticipated sales. It is unlikely that price expectations are the main cause of variations in levels of stocks. For instance, three producers said that they were forced to carry the barest margin of stock because of their tight liquidity position.

Most of the textile producers who were prepared to make a prediction expected wool prices to rise in the forthcoming year. But, despite this expectation, the wool stocks of most producers were normal in relation to their anticipated sales. This accords with the assertion of most Australian textile producers that their wool stocks are not affected, or substantially influenced, by their expectations of wool price movements. This, however, may not be so in large wool purchasing countries such as Japan.

8.3 THE FUNCTIONS, ORIGIN AND DEVELOPMENT OF COMMODITY FUTURES MARKETS

Functions of Commodity Futures Markets[4]

A simple definition of a commodity futures market is a place where forward contracts (futures) are traded [35]. Such a market incorporates the characteristics of both a forward and a cash market and in addition allows futures to be bought and sold in a way similar to that of share trading on stock exchanges. Because a futures market refines the existing cash market Tier defines a commodity futures market as:

> ... the medium for bringing together producers, merchants and processors who wish to buy or sell a commodity for delivery in a stated month - near or distant - without actually having to accept or make delivery unless they wish to do so [35, p. 71].

The operations of such a "medium", however, require certain requisites to be fulfilled. Among these Weisser [42], lists the following: (i) the commodity in question[5] must be subject to wide and continuous trading; (ii) it should be easily graded according to generally accepted and recognised standard grades; and (iii) the commodity should be subject to continuous and unpredictable price fluctuations.

According to Working [44], futures markets are especially prevalent in commodities associated with large price fluctuations arising from, among other things, unpredictable variations in production and other supply uncertainties. Such price fluctuations should be expected also [10]. These price fluctuations allow for the arrangement for protection from price risks and also provide opportunities to profit from speculation. A futures exchange caters for traders who wish to avoid price risks (hedgers) and who attempt to profit from futures trading activities (speculators). Both forms of trading activity will be examined shortly.

In addition to these requisites Goss [10] includes other conditions which he considers to be necessary for futures trading: (i) the commodity must be subject to rigidities in production which result in spot (cash) market commitments; (ii) the product must either be homogeneous or be capable of speculation at a standard grade (deviations from the specified grade must be

---

[4] A detailed account of the functions of futures markets is not undertaken here, but if desired can be obtained from Gold [9], and further economic aspects are covered in Hieronymus [19].

[5] Operations in futures markets have for a long time been common practice in the marketing of certain foodstuffs, fibres and metals.

measurable); (iii) the commodity, if it becomes the subject of a futures contract, must be deliverable and storage of that commodity must be possible; and (iv) liquidity in the form of sufficient funds must be available to enable market settlement. Some economists such as Gray [13] have indicated that even if these necessary conditions for futures trading are fulfilled, an active futures market is not guaranteed.[6]

According to Phillips "futures markets are the most advanced development in the accomplishment of economics in co-ordination over time" [30, p. 47]. In his view the greater liquidity which exists in futures markets compared to other forms of market organisation[7] facilitates this action with little or no delay and this decreases the uncertainty borne by individual traders.

A futures contract is a promise of the seller to deliver within a specified month and a promise of the buyer to take delivery of a standard quantity and quality of the commodity at an agreed price. Due to its homogeneous nature the commodity is readily adaptable to being traded in an exchange. The commodity exchange (futures market location) provides a central location where potential buyers and sellers make bids and offers for contracts covering delivery in various later months. At all times the number of obligations to accept delivery equals the number of obligations to make delivery - a trader wishing to discharge his obligations is assured always of an opposite number [44].

Trading of this kind can take place at any time up until the contract is due. Due to the centralised nature of the futures market and the speed and ease with which transactions (sales and purchases) can be made, the price of any sale in a given future (the delivery month in which trading takes place) is a "near perfect" reflection of supply and demand conditions existing at that instance of time for contracts of that particular future [25].

In most futures markets only a small fraction of contracts sold is actually delivered. Because nearly all participants are motivated by the desire to trade on price movements they liquidate by undertaking offsetting transactions. The buyer (seller) can liquidate his position in the futures market *prior to actual delivery* of the commodity by merely selling (buying) on the exchange, contracts of the same future - the second transaction offsets and cancels the previously existing contract.

The account of futures markets to this stage has dealt primarily with the general concept rather than the purpose of futures markets in terms of its participants. In regard to the purpose of futures markets, Working has defined a futures market as:

---

[6] For example, the contract may favour one side of the market, or one group of traders may be accustomed to a particular market arrangement and might be reluctant to change that tradition [10].

[7] A description of these other forms of market organisation and their comparison with futures markets in terms of liquidity is given by Bakken [2].

> ... trading conducted under special regulations and conventions, more restrictive than those applied to any other class of commodity transactions, which serve primarily to facilitate hedging and speculation by promoting exceptional convenience and economy of the transaction [44, p. 315].

Traders in futures markets can be classified primarily as hedgers or speculators although a situation known as arbitrage is referred to often.[8] This distinction between hedgers and speculators is made because these two groups benefit in different ways from their use of futures markets and accordingly each views the futures market in a different way [22].

The aim of the pure hedger according to Goss [10], is risk reduction.[9] He holds a position in the futures market that is opposite to the position he holds in the commodity (spot) market. Hedgers may comprise primary producers who desire greater income stability or manufacturers or merchants who are primarily concerned with the predictability of costs of their main input or their inventories [20]. There are two hedge positions that may be taken by the hedger; a short hedge or a long hedge. A short hedge involves the trader in selling futures initially and then purchasing them at a later data. In this way the short hedger protects his unsold stocks of future production against a price decrease in the commodity. An example of a short hedge is a wool grower who contracts to sell, on a greasy wool futures market, his entire anticipated wool clip. When his clip is auctioned he cancels his futures market position by purchasing futures.

A long hedger purchases futures contracts as protection against a cash price rise and hence covers his anticipated requirements of the commodity. Irrespective of the type of hedging undertaken, hedging is based on the assumption that risks are smaller with hedging than without [10].

A hedger, however, will not necessarily hedge all his stock at any one time. Some hedgers may keep only a part of their stock hedged. (Although hedging normally reduces risk it also increases cost.) Because of this some traders may be unwilling either to carry the whole risk of a given unhedged stock or to hedge it completely. They may think it would be to their advantage to hedge a certain part of it, thereby reducing risk to a tolerable level without reducing expected earning too much. In addition hedgers' operating practices vary because of their different circumstances, preferences, resources and abilities. Some traders who continuously buy and sell around the year may wish to carry only hedged stock (routine hedger) while others hedge only when

---

[8] Arbitrage is a situation taken in both the spot (physical) and futures market simultaneously, the aim being to gain a certain profit. The extent of this profit is normally known at the time of transaction.

[9] The concept of hedging has changed somewhat over the years. Good coverage of these changes can be found in Goss [10].

they expect a price fall but not when they expect a price rise (selective hedger).

Speculators (who can trade also short or long) aim on the other hand, to make uncertain profit from futures markets transactions. Such profits are based on the speculator's willingness to take risks and on his price expectation - which may be formed in terms of futures price, spot price or both. These price expectations may be formed precisely (different expected prices for specific futures dates with a clear idea of the possibility that each expected price will eventuate) or the speculator may have only a sketchy idea of the direction in which the price may change in the future [10]. Depending on his situation the speculator may be risk loving or risk-averting.

The speculator generally finds it advantageous to deal in futures contracts rather than either buying a quantity of a commodity at a current spot (cash) price and holding it in the hope of a rise in this price, or selling the commodity short by promising (in private negotiations with a private dealer) to deliver at a specified late date at a price which he expects to be above the spot price that will prevail at a later date [25]. As Hunt points out:

> The speculator whose aim is an uncertain profit is far more restricted to the nuances of the market than is the hedger who is merely reducing risk by maintaining a position in the futures market [22, p. 14].

If a speculator, who may be a dabbler or a large speculator (often making futures trading a full-time profession), takes up a bought position, it can be assumed that he expects prices to rise sufficiently within not too long a period to cover his transaction cost and provide an adequate return on money risked. If he sells it is assumed that he expects prices to fall sufficiently to give him a profit.

Trading on futures markets may depend to a large extent on hedging but in the opinion of Hodan "the workability of hedging depends on adequate well informed speculation and anticipation of futures price movements" [20, p. 7].

In relation to the important role played by speculation in futures market trading, Tier lists four market services provided by speculation "which greatly assist hedging activity" [36, p. 74]. The first of these services is the creation of a continuous market. By facilitating the existence of a continuous market speculation ensures that there is always a trader willing to take an opposite position to a buyer or seller and thereby enables transactions to take place at any time. The second service is that speculation reduces the prices effect in a market. That is, the difference between buyers' and sellers' price is reduced because speculators allow for frequent purchasing and selling which in a competitive market narrows the margin between buyers' and sellers' price.

The third service performed by speculation is that it facilitates risk transfer. Risks that would, in the absence of speculators, be the burden of stock holders "can be largely shifted to the shoulders of speculators by the process of hedging" [35, p. 75]. This can result in a cost reduction to "middleman" and can be reflected in lower prices to both consumers and producers. Finally, speculation can assist with price formation in that a large number of speculators operating in continuous competition ensures that prices reflect current knowledge in an accurate manner.

Origin and Development of Commodity Futures Markets

The first recorded case of organised futures trading occurred in Japan in the 1900s when wealthy landowners and feudal lords sold warehouse receipts against agricultural goods stored in warehouses. Merchants bought these tickets in anticipation of their projected needs.

The modern futures markets grew out of the time-contracts for grains at the Chicago Board of Trade in 1851. The first time-contract authorised the determination of 3000 bushels of corn to be made in June at a price one cent per bushel below the March price [23]. The commodity futures markets for grains have been and probably are still the most important [20].

Other early commodity futures exchanges were the New York Cotton Exchange (1870), the New Orleans Cotton Exchange (1870), the New York Coffee Exchange (1885), the Antwerp Wool Tops Futures Exchange (1887) and the Roubaix-Tourcoing Wool Tops Futures Exchange (1888). These early futures markets were considered primarily as delivery markets. Contracts to deliver would be undertaken and fulfilled to a large extent by deliveries [15]. In the view of Gray and Rutledge:

> Probably not the role, and certainly not the extent of hedging as it presently emerged was contemplated when the markets were organised. They were organised to facilitate the existing merchandising trade by providing uniform rules governing the transactions ... standard of grade, quantities and delivery terms [15, pp. 58-59].

However, Working [44] believes that the provision of hedging facilities was recognised early to be an advantage of futures markets. Once futures markets began to operate for Chicago grain, the demand for a means of hedging commercial risks was strong and persistent enough to permit this unconventional form of trade to overcome conservative opposition and the stormy period of experimentation required to give it a firm hold.

Futures markets were in fact better able to facilitate hedging than to direct merchandising. From their merchandising beginning they came to complement the existing trade by enabling futures prices to be established, and to which in addition, followed the other advantages of modern futures markets; namely, standardisation of the commodity, forward buying and selling for immediate or late delivery at negotiated prices of commodity lots specific as to location, grade, quantity, time and form of delivery.[10]

[10] A much more comprehensive analysis of the evolutionary aspects of future trading is undertaken by Gray and Rutledge [15]. See also Irwin [23].

Today, the U.S.A. is the country with the most comprehensive futures trading. Apart from Chicago, grain futures are traded also in Minneapolis and Kansas City. Important futures markets also exist in the U.S.A. for wheat, corn, sorghum, soy bean oil, rye, eggs, potatoes, cotton, cottonseed oil, lard, flaxseed, wool tops and greasy wool.

Other grain futures markets exist in Liverpool in the U.K., Winnepeg in Canada, Buenos Aires in Argentina and Rotterdam in The Netherlands. The leading cotton futures market is in New York but others exist in Chicago and New Orleans while an important cotton futures market is located also in Liverpool in the U.K.

Apart from the U.S.A., wool futures markets exist in Europe, U.K., Japan and Australia. As already indicated, wool tops futures markets were established and have been operating in Antwerp (Netherlands) since 1867 and in Roubaix-Tourcoing (France) since 1888. The New York Wool Tops Futures Market has been operating since 1931 and a greasy wool futures market has been operating in the same city since 1941. In London, futures markets in wool tops and greasy wool have operated since 1953 and 1963 respectively. In Japan worsted yarn futures markets were established in Nagoya (1951) and Osaka and Tokyo (1953) while most recently in Australia a greasy wool futures market (the Sydney Greasy Wool Futures Exchange) was established in Sydney in 1960.

In addition to all these various futures markets, important futures markets for many other staple commodities operate in Japan, and European countries such as France, U.K., Belgium and the Netherlands [40].

8.4 THE USE AND EFFECTIVENESS OF WOOL FUTURES MARKETS

The wool futures market primarily considered in this section is the Sydney wool futures market although references are made to other wool futures markets.[11]

Use of Wool Futures Markets

Australia is the largest producer of merino wool in the world and such wool in greasy form is highly suited to futures trading because it satisfies the necessary conditions listed in Section 8.3, namely, it is subject to considerable unpredictable price fluctuations, widely and continuously traded, storable and capable of being satisfactorily graded. In addition to this, greasy wool has been sold at open auctions, and until the establishment of the Australian Wool Board such markets were free from significant government interference.

---

[11] A description of the London wool top future market is contained in Brothwell [5], and Gutman and Duffin [16].

The Sydney futures market (Sydney Futures Exchange Limited)[12] began operation in May 1960. Previously Australian wool growers and processors had to use the London Wool Terminal Market but this was to their disadvantage because apart from the possibility of exchange rate alterations, the situation was exacerbated by the requirement of 12 months permission to transfer funds to the London Wool Terminal Market for hedging, and such permission was issued only by the Commonwealth Reserve Bank.

The Sydney Exchange evolved in part due to restrictions on capital movements out of Australia. The way was made easier in 1959 when the Commonwealth Treasury agreed to permit payments of futures contracts undertaken on an Australian wool futures market to be made to a non-resident in his particular country's currency.

A futures market located in Australia and providing for delivery was considered desirable by Australian wool growers, merchants, processors and brokers. The move for the establishment of the Sydney futures market came mainly from those who were most likely to sell wool forward and long hedgers (topmakers, spinners, weavers) than from those who were likely to hold actual stocks of wool (wool growers). The majority of hedgers at the Sydney Exchange seems to be long rather than short and there appears to have been little hedging by growers [43]. However, at least until recently there has been little pure speculative trading on the Sydney Exchange [35].

The establishment of the Exchange provided wool growers, merchants and manufacturers with hedging facilities against wool price fluctuations but speculators also have, and do use, the Exchange and as a consequence its functions have been given more utility. According to Hodan:

> Had there been no speculators, the operator who wanted to open or close a hedge would have had the difficult task of locating another *bona fide* hedger wanting to do just the opposite [21, p. 10].

A Sydney futures contract is for the greasy wool equivalent of 1500 kg clean weight of wool which is approximately 17 bales. Deliverable wools range from the 23 micron good topmaking medium length micron fleece wool up to the 19 micron average spinners and best topmaking warp merino fleece wool.

Eight types of contracts are traded simultaneously on the Sydney Exchange. Each contract is distinguished by name and it is possible for a contract to be satisfied by a trader if he takes an opposite position to the one he holds in the spot market. An example is the case of a trader who sells a contract to deliver

---

[12] The Sydney wool futures market was originally called the Sydney Greasy Wool Futures Exchange Ltd., but this was changed in 1972 to the Sydney Futures Exchange Ltd. This indicated the willingness of the Exchange to consider other forms of futures trading in addition to greasy wool [21].

wool deciding to close that transaction by buying a futures contract to accept 1500 kilograms of wool (the equivalent of one futures contract).

No futures contract is recognised as valid unless it has been registered with a clearing house. Clearing houses co-ordinate futures transactions, provide profit and loss settlements and supervise deliveries of contracts. Clearing facilities are provided by the Australian Development Clearing House Limited.

All floor members of the Sydney Exchange are required to register contracts in their own names. In 1974 the Sydney Exchange had 11 floor members and 51 Australian associate members together with 56 overseas associate members. A Board of Directors representing floor and associate members administered the Exchange and a committee of floor operators maintains an up-to-date set of trading rules. Only floor members may actually trade in futures on the Exchange floor and the number of floor members can be changed by a majority vote of the existing floor members.

Trading in wool futures is carried out only on the Exchange floor where bids and offers for the month open for trading are exchanged by the traders. Futures prices are quoted daily on the Sydney Exchange and at the end of each trading period closing quotations are declared for each month. Contracts can be registered only after such contracts have been traded openly ("by open outcry") as this ensures that prices are determined by all the available demand and supply. No further detail of the Sydney futures markets' operations is given but indepth coverage can be found elsewhere [14], [21], [30], [33], [42].

By the way, it is interesting to compare delivery on the Sydney Exchange with earlier contentions that little delivery takes place in commodity futures markets. On the Sydney Exchange less than two per cent of futures contracts are actually delivered.

Effectiveness of Wool Futures Markets

Several writers have considered the effectiveness of wool futures markets in fulfilling the objectives of their traders. Studies of the London and Sydney futures markets [16], [36], [42], using mathematical models based on hypothetical futures trading routines on these two markets have shown that the practice of continually hedging on the Sydney Exchange is less costly (including transaction costs) than on the London market. On a more general note, Phillips states that "... economics of co-ordination in time achieved by futures markets result in lower storage and processing costs" [30, p. 61].

Perhaps a more comprehensive indication of the effectiveness of wool futures markets is to test their efficiency. According to Praetz an efficient market is "a market in which prices fully reflect all available information" [31, p. 240]. From his four statistical tests on price changes on the Sydney Exchange Praetz obtained results which were consistent with the view that the Sydney wool futures market is efficient in using all available information. He notes that:

> The exact reason why the Sydney wool futures market seems to be efficient appears to be in the presence of very well informed traders who all agree on the month of all current and expected future information covering the market. Obviously the market is not fully efficient but there appear to be enough informed participants to reduce the size of any imperfections which occur to alleviate the size of transaction costs [31, p. 248].

Another view of efficiency is associated with the question of whether futures markets improve the allocation of resources through time [30]. According to Phillips a question such as this "is critical because the economies achieved in storage and processing may be more than outweighed by inefficiencies in allocation through time or a result of inefficient speculation" [30, p. 60]. He contends that the essential element in relation to futures market efficiency is whether there is more commodity price stability with the existence of a futures market than without.

Empirical tests have been carried out comparing price variability of a commodity before and after the establishment of a futures market. Such tests have, in general, tended to support the view that futures markets do reduce price variability [3], [29], [41], [45]. Specifically in regard to wool, Blau [3], found that the introduction of futures trading in the wool market reduced between-year wool price variability.

The evidence is by no means conclusive at this stage, however, and Gray and Rutledge [15], agree with Working that evidence about the role of futures markets in reducing price variability is:

> ... conclusive on only one point, namely that the average amount of seasonal variation in the price tends to be less in the presence of a futures market than in its absence [15, p. 44].

However, in so far as futures markets do reduce price variability they may increase the efficiency of production. Lower price variability means that producers may make more accurate estimates of futures prices and can thus allocate their resources in a more efficient manner [30].

The Sydney Wool Futures Market and Australian Textile Firms: Survey Results

In our supplementary survey [38], we investigated the extent to which Australian textile manufacturers used the Sydney wool futures market and their reasons for using or not using it. Manufacturers were asked whether or not the firm used or had used the wool futures market. All respondents to the survey (41) answered this question and, of these, 32 (78 per cent) indicated that they had not or did not use the Exchange, seven being yarn and fabric manufacturers, eight being yarn manufacturers and 17 being fabric manufacturers. Only nine of these 41 (22 per cent) indicated that they had used, or did use, the wool futures Exchange; three were yarn and fabric manufacturers, two were yarn manufacturers, one was a fabric manufacturer and three were top-makers.

The relative frequency of use of the wool futures Exchange, was greatest for topmakers, less frequent for yarn makers and least frequent for fabric manufacturers. Speculation in wool futures seems to be most pronounced in the earliest stages of wool processing.

Manufacturers' answers to this question can be related to their percentage use of wool in production. For simplicity, four categories of wool use are considered.

Category	Wool Content (by weight) in production
Low wool usage	1 - 30 per cent
Intermediate wool usage	40 - 79 per cent
High wool usage	80 - 99 per cent
Total wool usage	100 per cent

The relationship between manufacturers' relative dependence on wool in production and their use of the wool futures market is shown in Table 8.4. It is evident that the majority of manufacturers using wool did not use (or had not used) the wool futures market. Although not depicted in Table 8.4, the survey indicated that those manufacturers using only man-made fibres or cotton did not use (or had not used) the wool futures Exchange.

TABLE 8.4

Percentage of Wool Used in Production by Textile Producers Using Wool Related to Their Use of the Wool Futures Market

	Firms using the wool futures Exchange (now or earlier)					Firms not using the wool futures Exchange (now or earlier)			
Percentage of wool fibre usage	Yarn & fabric	Yarn	Fabric	Top-makers	Total	Yarn & fabric	Yarn	Fabric	Total
Low wool usage	-	1	1	-	2	1	2	6	9
Intermediate wool usage	1	1	-	-	2	2	1	-	3
High wool usage	2	-	-	-	2	2	2	2	6
Total wool usage	-	-	-	3	3	-	-	-	-
Total	3	2	1	3	9	5	5	8	18

Source: Supplementary survey.

Use or non-use of the wool futures market: Textile manufacturers participating in our supplementary survey were asked to state reasons for their use or non-use of the wool futures market.

Reasons for use: The most frequent reason put forward by textile manufacturers was that the Exchange allowed them to hedge against raw wool price rises. Two yarn and fabric manufacturers and three topmakers gave this as their reason. The three topmakers indicated that the wool futures market allowed them to hedge their yearly wool requirements against their forward sold contracts to manufacturers. One topmaker noted that it also used the Exchange to speculate. This accords with the contention of Durbin who says that:

> In practice, merchants and/or processors in wool feel free to adopt both speculative or a hedged position [on the wool futures market] and will do so in the light of information that is available to them [8, p. 73].

Through the use of hypothetical hedging transactions on the Sydney futures market Snape [33] found that price risks were reduced by hedging. Weisser also holds the view that although hedging on the Sydney futures market "does increase merchants and manufacturers costs ... it also reduces their risks" [42, p. 51]. He goes on to say that "by forward buying the manufacturer is relieved of the task of storing wool as well as of the risk of a price rise ..." [42, p. 51].

As an interesting empirical comparison Medcalfe [28], in his hedging model based on the Sydney Futures Exchange for the period 1972-1974 (a period of considerable wool price fluctuation) found among other things that routine hedgers (whether a long hedger or a short hedger) would have substantially offset the risk due to wool price variations in the spot market, if they had traded in futures. In testing a series of hypothetical hedging and speculative trading programmes on the Sydney futures market for the period January 1965 to December 1972 Goss [11] found that a majority (81 per cent) of the routine hedges taken were effective in risk reduction *ex post* and that the near future was a more effective hedging medium than the distant future.

Reasons for non-use: The principal reason given by textile firms for non-use of the wool futures market was that wool was not used in production. Fourteen manufacturers put forward this reason, ten being fabric manufacturers, three being yarn and fabric producers, and one being a yarn producer. Other reasons given were the following: (i) Three manufacturers, two yarn producers and one yarn and fabric manufacturer indicated that wool was only bought in response to market demand. (ii) One yarn and fabric producer said that instead of using the wool futures market, risks were averted by undertaking long-term contracts at a fixed price. (iii) Imported wool is used by one yarn and fabric manufacturer in the production of carpet and wool. (Futures quotations are not available on such imported wool.) (iv) Five fabric manufacturers indicated that they purchase their wool from spinners who may, or may not, use the wool futures market. (v) The length of contract was deemed to be of importance to one yarn and fabric manufacturer who noted that, as long as term contracts are not undertaken by him, the time involved in using the wool futures market is not justified. (vi) The amounts of wool used in production are too small to warrant participation in futures trading. However, the two firms (one a yarn producer, the other a fabric producer) who gave this as their reason did use the quotations of the wool futures market as an indicator of future trends, not only in wool prices but in the textile industry as a whole. As an interesting comparison, Durbin [3], found that European topmakers have used the Sydney Greasy Wool Futures Exchange also as an indicator of possible trends in the primary markets. He added, however, that not a lot of weight was placed on this as a means of forecasting.

Future use of the wool futures market: Thirty-eight manufacturers indicated whether they expected to use the wool futures market in the future. Thirty (79 per cent) said that they did not expect to use the market, and eight (21 per cent) indicated that they expected to use the wool futures market.

Reasons for expected use in the future:[13] Textile manufacturers advanced three reasons as to why they expected to use the wool futures market in the future: (i) Two producers indicated that they intend to use the wool futures market in the future, although they had previously not used it, because their use of wool was growing. One of these producers was a yarn producer and the other was a fabric manufacturer and both were low users of wool and high users of man-made fibres. (ii) Three firms, two fabric manufacturers and one yarn and fabric manufacturer, said that they intended to use the market but qualified their answers - their use would depend on prices quoted, customer demand for wool and the continued upward trend in the wool market. Similar results were recorded by Goss [12], who undertook a survey of traders on the Sydney Greasy Wool Futures Exchange between January and August 1970 (a sample of 25 was used of which seven were textile manufacturers or manufacturing merchants). Among other things, he found that production requirements (derived mainly from customer demand) were the main factors limiting the extent of their dealings on the wool futures market. (iii) The three topmakers put forward the same reasons as were mentioned in the last section but added that the degree of future use would depend to a large extent on the premium (discount) shown in futures operation - large forward premiums indicated expected wool price rises and it would be advisable to deal in futures under these conditions. The converse also holds. This is in line with Weisser who points out that:

> The discount, if any, if the forward price on the current auction price would be an incentive to forward buying, unless in the opinion of the manufacturer a decline in price is imminent which would exceed the discount offered by the forward seller. The forward seller himself must be

---

[13] Only reasons that differ from those given in *Reasons for use* in the last sub-section are given here.

> assumed to have precisely that view of the market, for otherwise the transaction would offer him no prospect of profit [42, p. 51].

As already noted, all topmakers in our supplementary survey used the Sydney Exchange to hedge against wool price rises. Durbin found from his survey of European topmakers that:

> ... with incomplete information, topmaking firms face situations involving risks. They may seek further information to reduce the level of risks or hedge their operations and therefore transfer a proportion of their risks to someone else [8, p. 54].

Reasons for expected non-use in the future: Two textile manufacturers, one a yarn and fabric manufacturer, the other a yarn manufacturer, had decided to discontinue their use of the wool futures market. Both indicated that they encountered difficulties because futures' quotations are based on the 21 micron wool and other wool types may not vary in price in the same manner as 21 micron. In particular they were users of coarse wool and found that prices for coarse wools have a varying relationship to that of 21 micron wools.

In addition to this reason, two producers, one yarn and fabric manufacturer and one yarn manufacturer, declared that it is not company policy to speculate. Their wool purchases are governed entirely by customer demands. In his survey, Goss [12], also found that company policy, as well as customer demands, were factors limiting the extent of dealings of traders on the Sydney Greasy Wool Futures Exchange.

The effectiveness of the Sydney wool futures market as means for avoiding risks associated with wool price fluctuations: Twenty-one manufacturers responded to the question of whether or not the wool futures market enables them to avoid the risks associated with fluctuations in wool prices. Of these 21, 12 (57 per cent) indicated t that the wool futures market did not *adequately* enable them to avoid risks associated with wool price fluctuations, while seven producers (33 per cent) said that the wool futures market did *adequately* enable them to avoid risks. Similar results were obtained from a previous survey of Australian textile manufacturers [7].[14]

Of the 12 producers from our supplementary survey who said the wool futures market did not allow risk avoidance, four were yarn and fabric manufacturers, five were yarn manufacturers, one was a fabric manufacturer and two were topmakers. The relative breakup varied slightly for these seven producers who said that the wool futures market did allow adequate risk avoidance; three were yarn and fabric manufacturers, two were fabric manufacturers, one was a yarn producer and one was a topmaker.

---

[14] This report is based on a survey of the Australian wool textile industry undertaken in late 1959 - during a period of considerable wool price instability. In that survey 30 per cent of the respondent textile manufacturers indicated that futures trading was effective in reducing risk associated with unstable wool prices.

Reasons given by producers claiming that the wool futures market adequately enables risk avoidance: Three reasons were put forward by textile producers: (i) Two topmakers said that the risk can be adequately avoided but not completely avoided. One also noted that futures quotations often move erratically in relation to the physical wool markets and, therefore, the avoidance of risk sometimes involves taking risks. (ii) One fabric manufacturer using high proportions of wool in its production said that the wool futures market had, in the past, enabled it to avoid risk adequately but this is no longer so because of government intervention and the disruptive interference of large multinationals. These two factors according to this manufacturer, have made rational estimation of wool price movements very difficult. (iii) One yarn manufacturer using high proportions of wool in its production indicated that the wool futures market shows a pattern of trends that are usually proven to be correct. This manufacturer, however, does not trade on the wool futures market but rather uses prices in the futures market for predictive purposes.

Reasons given by producers claiming that the wool futures market does not enable risk avoidance: Three reasons were put forward here: (i) Two yarn and fabric producers said that a price differential exists between carding and combing wool and the wool futures market does not adequately reflect future prices for carding wool. (ii) Certain wools have a varying price relationship to the wools of futures contracts on the wool futures market. One yarn and fabric manufacturer indicated that its wool requirements were primarily in the 23 to 30 micron range and reliable futures quotations could not be obtained. (iii) Wool futures prices and auction prices are not necessarily closely related - in the opinion of one topmaker, futures prices after time discount do not necessarily equal the actual price of greasy wool at auctions and he concluded that the wool futures market does not, therefore, enable risks to be avoided but rather minimises risks.

Views on whether the wool futures market is a useful stabilising agent in the marketing of wool: There were 20 respondents to the question of whether the wool futures market plays a useful stabilising role in the marketing of wool. Ten (50 per cent) indicated that the wool futures market did play a useful stabilising role in the marketing of wool and nine (45 per cent) said that it did not. The remaining producer was ambivalent in his answer.

Reasons given by producers claiming that the wool futures market plays a useful stabilising role in wool marketing: Most textile manufacturers who believed that the futures Exchange plays a stabilising role in the marketing of wool qualified their answers. The reasons given were the following: (i) One yarn and fabric producer said that the wool futures market is of some value because it gives an indication of expected future price levels by balancing normal supply and demand influences, but this, it added, can be disturbed by speculative trading. (ii) Two manufacturers, one producing yarn, the other producing fabric, claimed that the introduction of a floating (flexible) reserve price by the Australian Wool Corporation has made the stabilising role of the wool futures market less effective. (iii) One topmaker said that the wool futures market provides the facility for traders to purchase wool requirements at a time and place appropriate to their individual needs and consequently purchasing can be spread over a period and "pressure points" minimised. In addition, this manufacturer claimed that the wool futures market reflects a calculated assessment on the part of many interested parties in the wool textile industry, of the present and future prices of wool based on known factors. (iv) One yarn and fabric manufacturer said that the wool futures market aids wool marketing in the sense that it gives an early indication of a changed relationship between supply and demand - prices change in response to supply and demand.

Reasons given by producers claiming that the futures market does not play a useful stabilising role in wool marketing: There were three reasons put forward by textile producers: (i) One yarn and fabric manufacturer said that futures quotations are based on 21 micron wool and consequently the use of the Exchange as a stabilising agent is limited because prices of wool other than 21 micron move differently to that of the 21 micron. (ii) Some manufacturers said that wool prices are subject to violent fluctuations and that this is evidence that the wool futures market does not exert a stabilising influence in wool marketing. Although the wool futures market may be able to exert a stabilising influence it does not do so at present - this view was expressed by a yarn and fabric manufacturer using a high proportion of wool in his production and supported by a topmaker. The topmaker said that, although the wool futures market plays a useful role in wool marketing, it does not perform the functions of a stabilising agent. (iii) While the wool futures market once performed a stabilising role it no longer does so in the view of one fabric manufacturer because prices have become more uncertain. Uncertainty has increased in his view because of government intervention in the marketing of wool and because of the presence of activity by multinational companies. Adding to this, one yarn and fabric manufacturer said that the wool futures market was open to speculation on a large scale, particularly from the Japanese and that this diminished the stabilising role which the market could play.

It is clear that a majority of textile manufacturers have not used and do not use the wool futures Exchange. Even a majority of producers using wool in their production do not use the Exchange. In addition, the relative frequency of use of the Exchange is much higher in the earlier stages of textile processing than in the later stages - all the topmakers responding to the survey used the wool futures Exchange, but few yarn makers took advantage of it and even fewer fabric makers were involved in trading on the Exchange.

The composition of textile manufacturers trading in wool futures on the Exchange is not expected to alter in the foreseeable future. On the whole, those producers currently trading in futures expect to continue to do so and those not trading in futures expect to continue not to trade on the Exchange.

Most of the textile manufacturers surveyed believe that the futures Exchange does not adequately enable them to avoid risks associated with fluctuations in wool prices. However, a majority, but not an overwhelming majority of respondents do believe that the wool futures Exchange plays a useful stabilising role in the marketing of wool.

8.5 A GUARANTEED PRICE FOR WOOL

The last section has indicated that textile manufacturers with forward commitments such as yarn makers, weavers and knitters can obtain a certain measure of protection against adverse wool price fluctuations by trading in wool futures markets. Other methods can also be utilised by textile manufacturers for this purpose. These include price averaging of stocks and the off-setting of all forward sales by continued forward purchasing. However, in the view of the Australian Wool Corporation [1] none of these measures allows the textile producer to perfectly hedge his risks against possible wool price fluctuations. In its opinion "the predominant view among processors is that they would prefer to operate in a more stable trading environment" [1, p.66].

The establishment of a government guaranteed forward price for wool may help to bring about such stability. D.G. Johnson [24], has pointed out that guaranteed forward prices for agricultural products may result in increased stability. It was our intention to find out from our supplementary survey whether textile processors wished to operate in the more stable and certain environment which might result from the introduction of a forward price for wool.

Attitudes to a Guaranteed Price for Wool: Survey Results

Textile manufacturers were asked whether they would prefer, in principle, a government guaranteed annual wool price to the existing market situation for wool. Twenty-five manufacturers responded of which 22 (88 per cent) favoured a guaranteed price and three (12 per cent) did not favour the proposal. Of the 22 favouring the proposal, seven were yarn and fabric manufacturers, six were yarn manufacturers, eight were fabric manufacturers and one was a topmaker. Of the three that did not favour the proposal, two were topmakers and one was a yarn and fabric manufacturer.

The majority of topmakers were not in favour of a guaranteed price for wool, presumably because it might reduce their income as risk takers and speculators in the industry. Another interesting result is that processors earliest in the processing chain are less likely to favour a guaranteed price than those further down the processing chain. Apparently, the later the stage of the production process, the less likely is the processor to be interested in speculation or risk taking.

Reasons for favouring the guaranteed price proposal: The most frequent reason given by textile processors for favouring the guaranteed price proposal was because it would help to minimise the risks associated with uncertain and unstable wool prices. Twelve textile producers gave this reason and some also felt that their use of wool would rise because the guaranteed price proposal would reduce fluctuations in wool prices and help stabilise the market. One of the respondents specifically stated that a policy of this kind would be useful in counteracting the stable market for synthetics and could enhance competition with man-made fibres.

Other reasons put forward included the following: (i) Three firms, a yarn and fabric manufacturer, a yarn producer and a fabric producer, said that a guaranteed price scheme would facilitate planning and control of costs and cash flows because much of the uncertainty associated with wool price fluctuations would be removed. The yarn producer indicated that the likelihood of giving "firm" forward orders of up to 12 months would be enhanced under such a scheme. However, some of these views were qualified. The fabric manufacturer noted that although a scheme of this nature would remove a large amount of uncertainty from forward costing, the guaranteed price level would need to be applied to yarn imports or to the wool content in imported yarn. In addition, this firm stressed that for the scheme to be effective, it would need to be worldwide. (ii) One manufacturer said that such a scheme would allow the introduction of a system whereby wool could be successfully marketed in the long-term. The yarn and fabric manufacturer who put this reason forward stressed that trading terms for wool should be similar to those for synthetic fibres. He indicated that because modern spinning machinery is highly selective, failure to comply with this could lead to a drastic reduction in the demand for wool.

Reasons for not favouring the guaranteed price proposal: One topmaker said that he would prefer a free market to the present system. The existing guaranteed minimum price for Australian wool put the company out of business for 12 months during the recent world textile recession when the top market was largely satisfied by purchases of non-Australian wool. Another topmaker stressed that it was not practical for one country to try and manipulate the price for a world commodity.

It is clear that a majority of textile manufacturers surveyed favoured the guaranteed price proposal. Risk minimisation, stability of market and the facilitation of planning and control of stocks were the main reasons put forward for this favourable response. The major objection to the guaranteed price for wool came from topmakers who presumably fear that such a scheme will adversely affect their operations as risk-takers and/or speculators.

REFERENCES

[1] Australian Wool Corporation, *The Marketing of Australian Wool*, Melbourne, 1973.

[2] BAKKEN, H.H., *Theory of Markets and Marketing*, MIMIR, Madison, 1952.

[3] BLAU, G., "Some Aspects of the Theory of Futures Trading", *Review of Economic Studies*, 12, 1-30, 1944-45.

[4] BRENNAN, M.J., "The Supply of Storage", *American Economic Review*, 48, 50-72, 1958.

[5] BROTHWELL, J.F., "The London Wool Futures Market", Appendix A in G.F. Rainnie (ed.), *The Woollen and Worsted Industry: An Economic Analysis*, Clarendon Press, Oxford, 1965.

[6] CARLAND, D.J., "A Model of the Japanese Wool Textile Industry", *19th Annual Conference, Australian Agricultural Economics Society*, La Trobe University, 1975.

[7] Department of Trade (Industries Div.) and Bureau of Agricultural Economics, *The Australian Wool Textile Industry and its Use of Wool and Other Fibres*, Canberra, 1961.

[8] DURBIN, S., *Short Term Price Formation in the Wool Market*, University of New England, Armidale, N.S.W., 1975.

[9] GOLD, C., *Modern Commodity Futures Trading*, Commodity Research Bureau Inc., New York, 1971.

[10] GOSS, B.A., *The Theory of Futures Trading*, Routledge and Kegan Paul, London, 1972.

[11] GOSS, B.A., "Trading on the Sydney Wool Futures Market: A Test of a Theory of Speculation at the Level of the Individual", *Australian Economic Papers*, 11, 182-202, 1972.

[12] GOSS, B.A., "Survey of Hedging Theory and Tests of Hedging Hypothesis on Sydney Greasy Wool Futures Exchange", *Fourth Conference of Economists*, Canberra, 1974.

[13] GRAY, R.W., "Why Does Futures Trading Succeed or Fail: An Analysis of Selected Commodities", *Futures Trading Seminar*, MIMIR, Madison, 1966.

[14] GRAY, R.W., *Wool Futures Trading in Australia - Further Prospects*, Research Bulletin No. 5, Dept. of Agricultural Economics, University of Sydney, 29-32, 1967.

[15] GRAY, R.W., and RUTLEDGE, D.J.S., "The Economics of Commodity Futures Markets: A Survey", *Review of Marketing and Agricultural Economics*, 39, 57-108, 1971.

[16] GUTMAN, G.O., and DUFFIN, B.R., "The London Wool Futures Market", *Quarterly Review of Agricultural Economics*, 8, 185-192, 1955.

[17] HAWTREY, R.G., "Mr Kaldor on the Forward Market", *Review of Economic Studies*, 7, 202-205, 1939-40.

[18] HICKS, J.R., *Value and Capital*, Oxford University Press, London, 1939.

[19] HIERONYMUS, T.A., *Economics of Futures Trading*, Commodity Research Bureau, Inc., New York, 1971.

[20] HODAN, M., "Fundamentals of Commodity Futures Trading", *Farm Policy*, 14, 3-9, 1974.

[21] HODAN, M., "Sydney Wool Futures Market", *Farm Policy*, 14, 10-12, 1974.

[22] HUNT, B.F., "Short Run Price Cycles in the Sydney Wool Futures Market", *Australian Journal of Agricultural Economics*, 18, 133-143, 1974.

[23] IRWIN, H.S., *Evolution of Futures Trading*, MIMIR, Madison, 1954.

[24] JOHNSON, D.G., *Forward Prices for Agriculture*, University of Chicago Press, Chicago, 1947.

[25] JOHNSON, L.L., "The Theory of Hedging and Speculation in Commodity Futures", *Review of Economic Studies*, 27, 139-51, 1960.

[26] KALDOR, N., "Speculation and Economic Stability", *Review of Economic Studies*, 7, 1-27, 1939-40.

[27] KEYNES, J.M., *A Treatise on Money*, Harcourt, New York, 1930.

[28] MEDCALFE, J.S., "A Note on the Cost and Effectiveness of Hedging Wool on the Sydney Futures Exchange", *Quarterly Review of Agricultural Economics*, 29, 42-49, 1976.

[29] PAVASAKAR, M.G., *Marketing of Cash Crops: Efficiency of Futures Trading*, Economic and Political Weekley, Bombay, 1971.

[30] PHILLIPS, J., "The Theory and Practice of Futures Trading", *Review of Marketing and Agricultural Economics*, 34, 43-63, 1968.

[31] PRAETZ, P.D., "Testing the Efficient Market Theory on the Sydney Wool Futures Exchange", *Australian Economic Papers*, 14, 240-249, 1975.

[32] SAMUELSON, P.A., "Proof that Properly Anticipated Prices Fluctuate Randomly", *Industrial Management Review*, 6, 41-49, 1965.

[33] SNAPE, R.H., "Price Relationships on the Sydney Wool Futures Market", *Economica*, 35, 169-178, 1968.

[34] STEIN, J.L., "The Simultaneous Determination of Spot and Futures Prices", *American Economic Review*, 51, 1012-1025, 1961.

[35] TIER, T.J., "Futures Speculation and the Sydney Greasy Wool Futures Market", *Quarterly Review of Agricultural Economics*, 23, 71-81, 1970.

[36] TIER, T.J., *A New Zealand Crossbred Wool Futures Market: Report to the New Zealand Wool Board - Wool Commission's Wool Marketing Study Group*, 1968.

[37] TISDELL, C.A., KEATING, G.R., and McDONALD, P.W., *A Survey of Australian Textile Manufacturers: A Short Report of Results for Participating Yarn and Fabric Manufacturers and Topmakers*, The University of Newcastle, N.S.W., 1975.

[38] TISDELL, C.A., and McDONALD, P.W., *Supplementary Survey of Australian Textile Manufacturers: A Report on Fluctuating Wool Prices, Stock Holdings and Use of Futures*, The University of Newcastle, N.S.W., 1976.

[39] TEWELES, R.J., and others, *The Commodity Futures Game*, McGraw-Hill, New York, 1974.

[40] United States Department of Commerce, *Inventories in the Textile Cycle*, Washington, 1961.

[41] VEIN NAIK, A.S., *Effects of Futures Trading on Prices*, Somoiya, Bombay, 1970.

[42] WEISSER, M., *The Sydney Wool Futures Market*, Wool Economic Research Report, No. 5, Bureau of Agricultural Economics, Canberra, 1963.

[43] WEISSER, M., "Cost and Benefit of Hedging Wool", *Quarterly Review of Agricultural Economics*, 15, 160-169, 1961.

[44] WORKING, H., "Futures Trading and Hedging", *American Economic Review*, 43, 314-343, 1953.

[45] WORKING, H., "Price Effects of Futures Trading", *Food Research Institute Studies*, 1, 185-220, 1960.

[46] WORKING, H., "New Concepts Concerning Futures Markets and Prices", *American Economic Review*, 52, 431-459, 1962.

APPENDIX TO CHAPTER 8

FIBRE AND TEXTILE ECONOMICS RESEARCH PROJECT

Supplementary Questionnaire

Kindly complete and return the questionnaire not later than May 31st, to

Mr P.W. McDonald,
Department of Economics,
UNIVERSITY OF NEWCASTLE N.S.W. 2308.

NAME OF FIRM_________________________________

NAME AND POSITION OF PERSON(S) COMPLETING THIS QUESTIONNAIRE ON BEHALF OF FIRM

____

____

SECTION A

1. Have you used or do you use the wool futures market?

YES ☐

NO ☐

WHY?____

____

____

____

____

2. Do you expect to use the wool futures market in the future?

YES ☐

NO ☐

WHY?____

____

____

____

____

SECTION B

*Questions 3 to 8 are primarily intended for firms using various proportions of wool in their production.

3. Do you believe that wool futures market adequately enables you to avoid the risks associated with fluctuations in wool prices?

YES [ ]

NO [ ]

WHY?---

----

4. Do you believe that the wool futures market plays a useful stabilising role in the marketing of wool?

YES [ ]

NO [ ]

WHY?---

----

5. Would you prefer, in principle, a guaranteed price for wool (guaranteed ahead by the government on an annual basis) to the existing market situation for wool?

YES [ ]

NO [ ]

WHY?---

----

6. Are your holdings of wool stocks (in various forms) influenced by your expectations of wool price movements?

   YES ☐

   NO ☐

   If yes, in what ways are your holdings influenced?

   If no, why?

   ----

7. Are your current wool stocks (in various forms), in relation to your anticipated sales

   (i) Higher than normal? ☐

   (ii) Lower than normal? ☐

   (iii) Normal? ☐

8. Do you expect wool prices in the next year to

   (i) Rise? ☐

   (ii) Fall? ☐

   (iii) Remain unchanged? ☐

CHAPTER 9

Wool Promotion: Strategic Considerations

9.1 INTRODUCTION

The wool industry has responded to increased competition from the man-made fibre industry by increasing the intensity of its promotional effort. This is most clearly seen by considering Australia's contribution to wool promotion as a percentage of the value of wool produced, a contribution which far exceeds that of any other country. This is graphed in Fig. 9.1.

A levy was imposed upon Australian woolgrowers from 1937 onwards to finance the promotion of wool but until the early 1960s the financial sum involved amounted to less than half a per cent of the gross value of Australian wool production.[1] During the early 1960s the International Wool Secretariat emphasised the need for greater promotional effort by the wool industry to counteract the increasing dangers posed to the industry by synthetic fibres. To meet these challenges, the International Wool Secretariat estimated that it needed a four-fold increase in funds from its member countries. Consequently, Australian woolgrowers approached the Federal Government for aid. Following the report in 1962 of the Wool Marketing Committee of Enquiry [28], (the Philp Committee) which strongly supported the desirability of increasing expenditure on wool promotion, the Australian Government provided matching assistance to growers on additional funds raised by them for promotion. Since the introduction of the subsidy for wool promotion, its basis has varied depending for instance upon economic circumstances. The relative contribution of the Government increased substantially in the early 1970s when wool prices were extremely depressed.[2]

The intensity of promotional effort has oscillated since 1964-65 because price fluctuations, in the main, have caused the value of

---

[1] For a review of the levy system see Donaldson [7] and Tisdell [21].

[2] Arguments for and against government assistance for wool promotion are set out in the Industries Assistance Commission's *Draft Report* [10]. Some further theory of relevance to this matter can be found in De Boer [6], Sturgess [20], and Tisdell [22], [24].

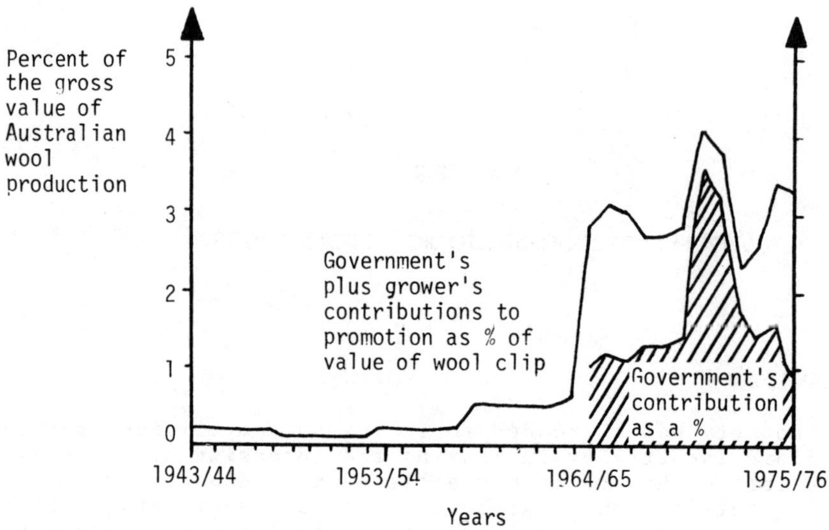

Fig. 9.1

Promotional Intensity of the Australian Wool Clip 1943-44 to 1975-76

Source: Based on figures in Table 2 of *Draft Report on Financing Promotion of Rural Products* [10].

the Australian wool clip to vary greatly and growers' contributions for promotion have been based on a relatively fixed percentage of this value. These fluctuations affect the total amount of funds available to bodies such as the International Wool Secretariat for promotional purposes and hamper their long-term planning.

The total value of the Australian government and growers' contribution to wool promotion since 1943-44 is graphed in Fig. 9.2. This sum is graphed at current prices and at 1966-67 constant prices based on the Consumer Price Index, an index which is readily available but of debatable relevance in this case. While the money sum for wool promotion has tended to rise the real sum has been falling since 1965-66.

In presenting its evidence to a recent enquiry of the Industries Assistance Commission, the International Wool Secretariat (I.W.S.) calculated a price index based on media costs in a number of countries and salaries and wages. Using this index, the I.W.S. was able to show that its real expenditure has declined since 1966-67. Between 1966-67 and 1975-76 the real expenditure of the I.W.S. fell by more than 30 per cent.

While the promotional intensity of the Australian wool clip has not fallen significantly and remains at a high level, real expenditure on wool promotion is falling. It is therefore necessary to reconsider the optimality of the present level of effort. A small

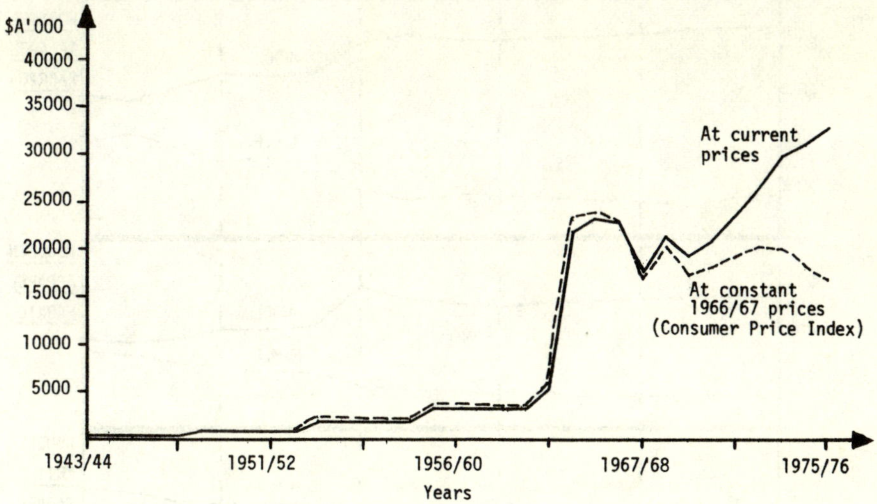

Fig. 9.2

Total Value of Australian Government and Growers' Contribution to Wool Promotion 1943-44 to 1975-76

Sources: Based in part upon figures in Table 2 of *Draft Report on Financing Promotion of Rural Products* [10], figures in *Yearbook of Australia 1974* [1] and *Quarterly Summary of Australian Statistics* [3].

start in this direction is made in this Chapter by considering the relevance of the Dorfman-Steiner theorem [8] and the Nerlove-Waugh theorem [16] for determining optimal promotion levels in the wool industry. Important theoretical questions about the promotion of wool mixtures are also considered.

9.2 WOOL CONSUMPTION IN VARIOUS END-USES AND SALES OF WOOL MIXTURES

Before considering these theoretical questions, it is interesting to digress and examine trends in the market share of wool in its main uses in the wool textile industry. Despite the high promotional intensity achieved in the wool industry from the early 1960s onwards, Figs 9.3 and 9.4 indicate that the proportion of wool used in all end-uses, with the exception of blankets, has declined. While reduction in the share of wool in the worsted and woollen apparel fabric markets has been slight, in carpets and double-knit fabric the fall has been extremely great because in part technological developments in carpet production and knitting have favoured man-made fibres.

Whereas relative consumption of wool has tended to decline in all main end-uses of wool, consumption of man-made fibres has risen. It is interesting to consider the extent to which this is

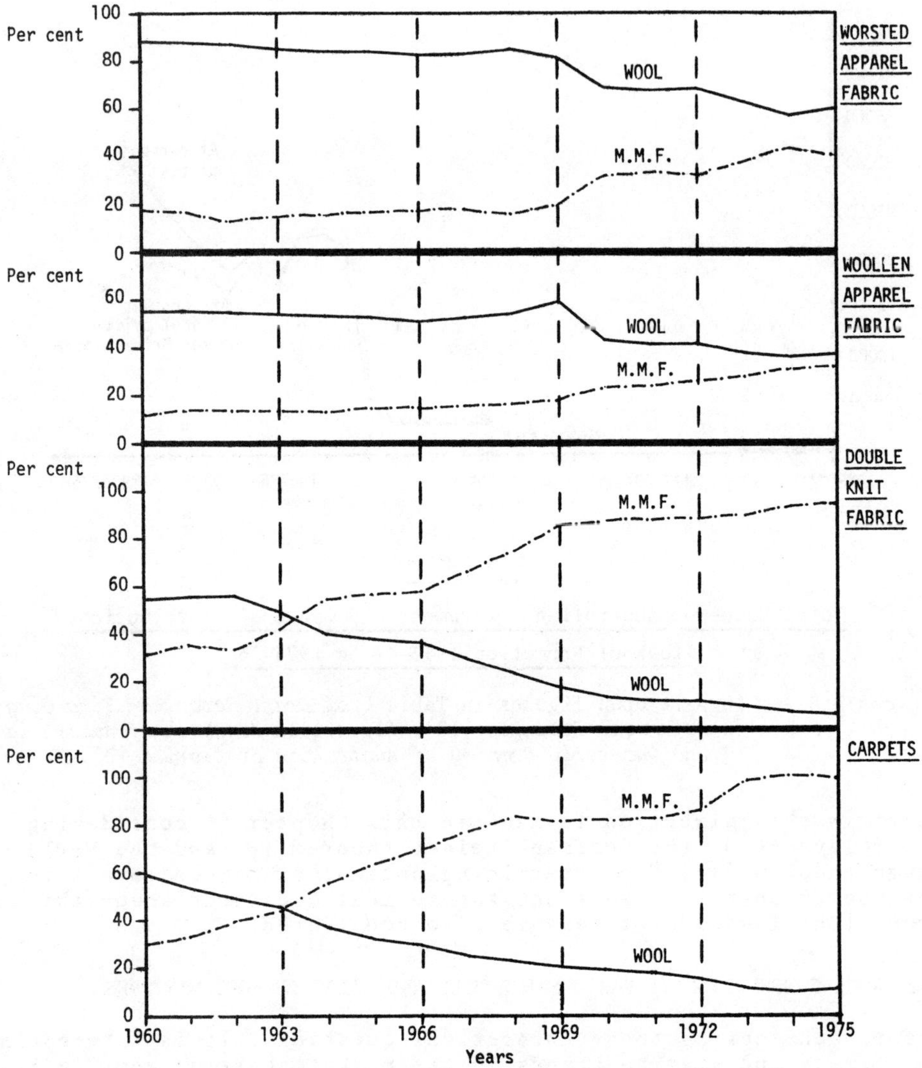

Fig. 9.3

Percentage of Wool and Man-Made Fibres Used to All Fibres Consumed[(a)] by Weight in End-Uses Indicated 1960 to 1975

Note: (a) Fibre consumption in the wool textile industry of eight producing countries - U.S.A., Japan, U.K., France, Italy, Belgium, Germany (F.R.), and The Netherlands.

Sources: Based on figures in Table AXXVI in *A Review of Developments in Man-Made Fibres 1960-1971* [18] and Table III.15 in *The Fibre Review 1975-76* [5]

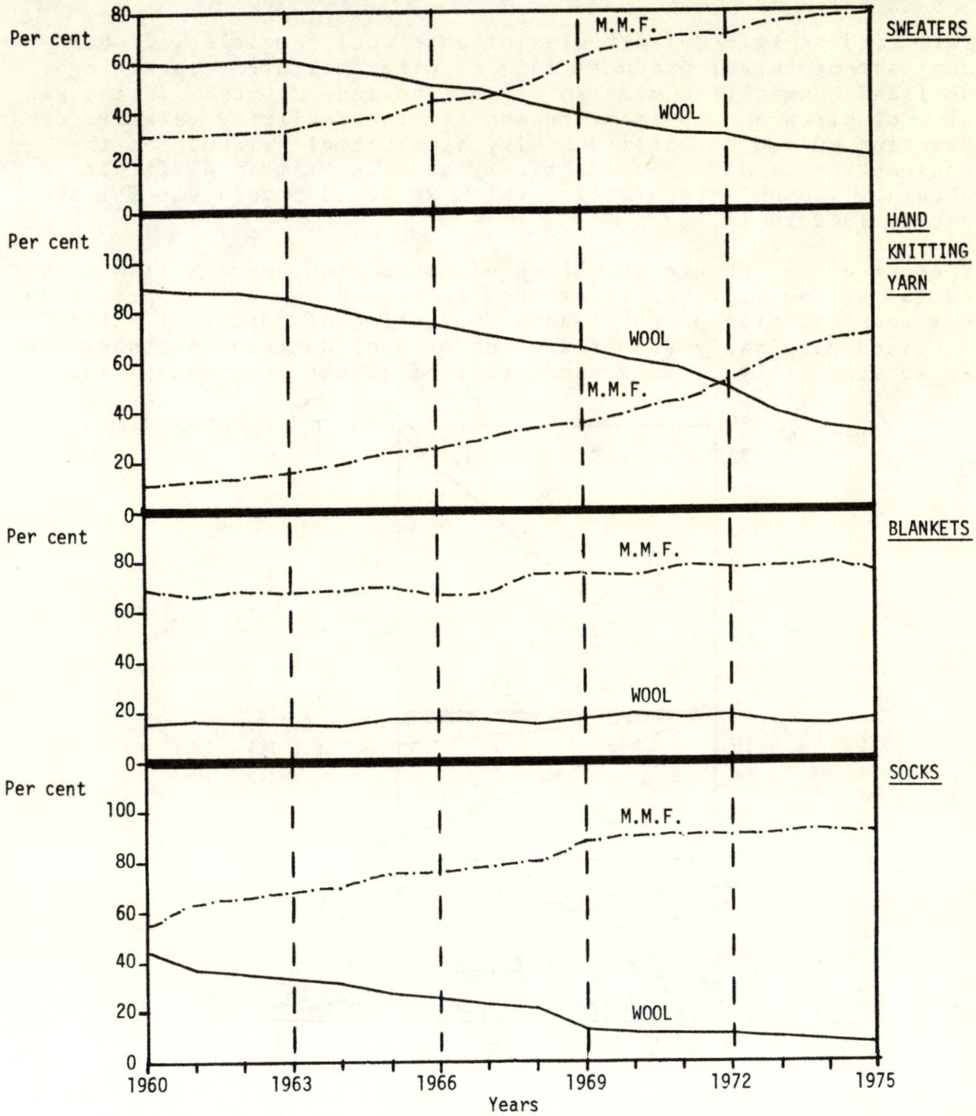

Fig. 9.4

Percentage of Wool and Man-Made Fibres to All Fibres Consumed by Weight in End-Uses Indicated 1960 to 1975

Note: (a) Fibre consumption in the wool textile industry of eight producing countries - U.S.A., Japan, U.K., France, Italy, Belgium, Germany (F.R.), and The Netherlands.

Sources: Based on figures in Table AXXVI in *A Review of Developments in Man-Made Fibres 1960-1971* [18] and Table III.15 in *The Fibre Review 1975-76* [5].

reflected in sales of articles of pure wool, articles of wool dominant mixtures, products made of other mixtures containing wool and commodities made solely of man-made fibres. Is the use of wool mixtures a primary reason for the declining relative consumption of wool? Unfortunately, statistical evidence on the consumption of different fabric types is extremely difficult to obtain. The only information which we could obtain was for the decade 1960 to 1970.[3]

The percentage change in shares of men's ready-made suits accounted for by fabric types is graphed in Fig. 9.5. In the period under consideration proportionate production of pure wool suits declined marginally and production of wool dominant suits declined substantially. The fastest rate of growth occurred in sales

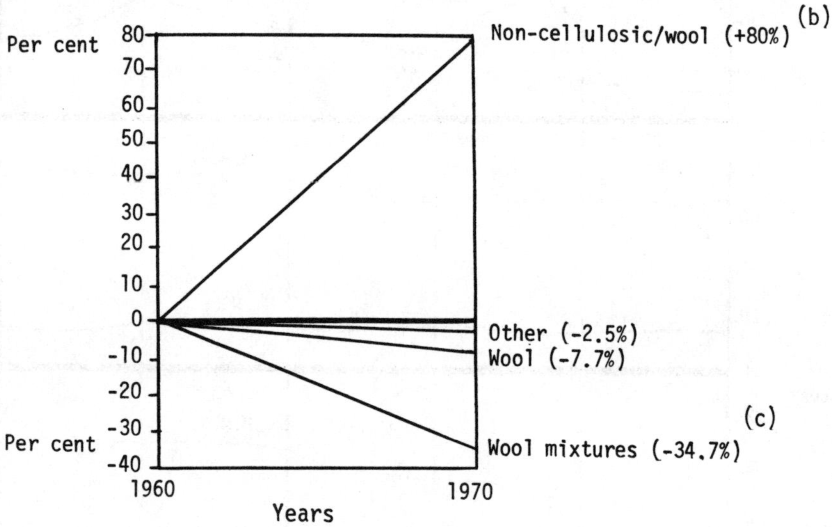

Fig. 9.5

Percentage Change in the 1970 Share of Men's Ready-Made Suits Accounted for by Fabric Type in 1970 compared to 1960[a]

Notes: (a) Covers eight producing countries - U.S.A., U.K., Japan, Fran ce, Germany (F.R.), Italy, Belgium, The Netherlands.

(b) Non-cellulosic dominant.

(c) Wool dominant.

Source: Based on statistics in Table A.XXX in *A Review of Developments in Man-Made Fibres 1960-1971* [18].

[3] We approached the International Wool Secretariat for further information but without success. The Industries Assistance Commission in its *Draft Report* [10] also complained about the unavailability of statistics dealing with this matter.

of non-cellulosic dominant wool blends which accounted for most of the growth in sales of ready-made suits. Sales of wool dominant mixtures declined absolutely and relatively in the period. During this period the sales of blends appear to have favoured sales of man-made fibres. However, it must be remembered that at that time the I.W.S. had not introduced Woolblendmark (this was introduced in 1971) and I.W.S. was not an active promoter of blends. In the light of new developments, it is unfortunate that later statistics are not available.

Statistics for the production of men's trousers for the period 1960 to 1970 reveal that absolute sales of wool trousers remained almost stationary and sales of trousers made of wool dominant blends declined. By contrast sales of non-cellulosic dominant wool blends increased to 2.5 times their 1960 sales and sales of trousers made entirely of non-cellulosics increased to almost seven times their 1960 sales. By 1970 over 40 per cent of all trousers were made of synthetic fibres. The percentage change in shares of the production accounted for by various fabric types is graphed in Fig. 9.6. Despite the falling sales of wool

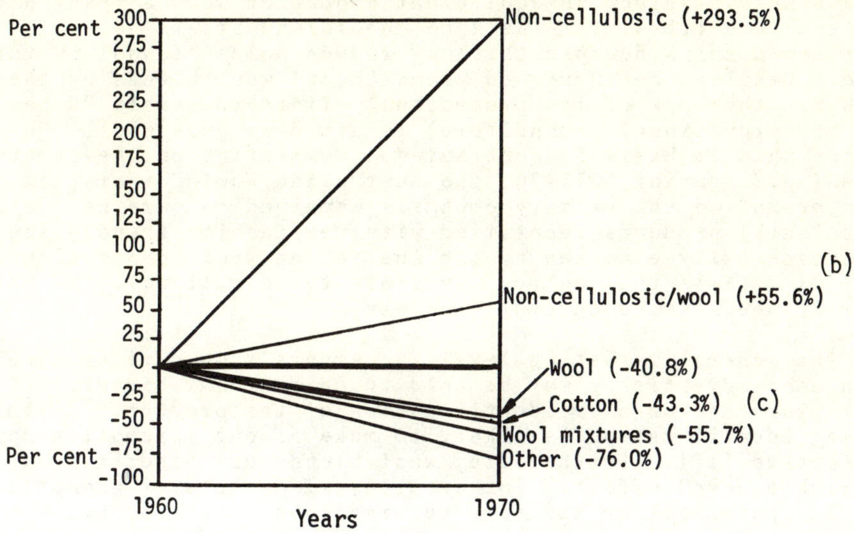

Fig. 9.6

Percentage Change in the Share of Men's Trousers Accounted for by Fabric Types in 1970 Compared to 1960[a]

Notes: (a) Covers eight producing countries - U.S.A., U.K., Japan, France, Germany (F.R.), Italy, Belgium, The Netherlands.

(b) Non-cellulosic dominant.

(c) Wool dominant.

Source: Based on statistics in Table A.XXXII in *A Review of Developments in Man-Made Fibres 1960-1971* [18].

dominant blends, wool consumption in this end-use still rose by 25 per cent, largely as a result of the increased quantity of wool consumed in the production of non-cellulosic dominant wool blends.

Clearly sales of wool mixtures have a bearing on the volume of wool sold. The policies which the wool industry should follow in promoting fibre mixtures containing a proportion of wool have been debated by the industry for a number of years and the wool industry now promotes wool mixtures to a limited extent. The I.W.S. has for some time given technical advice to textile producers using wool in their blends. In 1971 the I.W.S. introduced Woolblendmark to identify wool-rich blends and to provide a symbol to promote these. Woolblendmark is available to textile producers for wool-rich blends (those containing 60 per cent or more wool), for selected products only and subject to the maintenance of general quality standards by the user [12]. In particular Woolblendmark is only made available for the promotion of a blended product when the I.W.S. is convinced that the promotion of the product will result in a net increase in the demand for wool and not simply displace an equivalent amount of wool already used. In some countries, for instance, Woolblendmark is not available for men's suits because this may reduce sales of wool in this end-use. Despite the increased promotion of wool blends by the I.W.S., the bulk of its promotional effort (possibly 90 per cent of its promotional expenditure) is aimed at pure wool. Furthermore, this emphasis is continuing. Commenting on the strategy of the I.W.S. during 1975-76, the Australian Wool Corporation said: "In promotion the primary emphasis remained on pure new wool (Woolmark) products identified with new fashion trends, an approach fully endorsed by the Australian Wool Corporation" [4, p. 19]. While this policy may prove to be rational, the basis for it needs to be carefully examined.

At the general marketing level, an emphasis on pure new wool in consumer advertising may be held to be rational because this does not blur the image or identification of the product. A simple, clear identification is likely to make product promotion more effective [13]. Furthermore, wool blends may benefit by a "trickle-down" effect. For example, if a consumer cannot afford to buy pure wool or has some reservations about it, his preference for wool blends may be increased by an advertising campaign in favour of pure wool.

Again, the view has been expressed by some leaders in the wool industry that it may be unnecessary to promote wool mixtures because these will be promoted by manufacturers and processors of man-made fibres. Because mixtures have superior characteristics for some purposes [11] or are initially more acceptable to consumers, manufacturers of synthetic fibres may promote these in order to increase their sales of synthetics. The reliance which the wool industry can place on this factor is considered later.

The continuing policy importance of the promotion of blends is brought out by the following extract from the *Interim Annual Report 1975-76* of the Australian Wool Corporation:

> During the year a committee of senior members of the line and staff management [of IWS] reviewed in detail Woolmark and Woolblendmark activities. It was recommended that the Woolmark must remain the chief vehicle and focus of IWS promotional activities. The Woolmark programme is now 12 years old, and represents a considerable investment on behalf of woolgrowers. Its success is evident in the recognition and understanding of the symbol worldwide. Moreover, a high proportion of raw wool is used in 100 percent pure new wool merchandise. In 1974, for example, which was not the best year for wool, almost two-thirds of the virgin wool consumed in major IWS countries was used in products of this type.
>
> Given the continued key role of the Woolmark, however, the IWS has recognised that the textile environment is changing. The consumer protection movement - which in many ways has been so helpful to wool - has resulted in labelling legislation in many countries which has reduced the Woolmark's function as a guarantee of fibre content. Its role as a guarantee of quality, however, is likely to become even more important in the future.
>
> The Woolblendmark, sister symbol to the Woolmark, which is used to denote wool-rich blends, has also been successful in promoting a carefully restricted range of products, such as socks and men's trousers. Urgent studies are now being carried out on several products to establish whether their inclusion in the Woolblendmark campaign will result in a net gain in wool usage.
>
> The thorough review of the Woolmark and Woolblendmark programmes has reinforced the view that there is no case for swinging the major part of the IWS's promotional effort behind blends. What has emerged, however, is a somewhat more flexible policy, with built-in safeguards. This will strengthen the hand of the IWS in its efforts to expand markets for wool [4, pp. 20-21].

The extent that the wool industry should promote blends has been and is a contentious issue. In 1971, just after the introduction of Woolblendmark, Edwards and Trickett said:

> The decision to promote pure wool as a superior good and a differentiated product in opposition to other fibres was ill-advised, in view of trends in the fibres industry as a whole. Despite vast technological improvements in the properties of wool fibres in recent years, wool garments did not possess the vivid dyeing and "easy care" washing properties of other fibres, and consequently lost ground in the popular and everyday items of apparel. There is no doubt that wool retained its quality image in fashion garments - but the fashion market is perhaps the most volatile of all apparel markets, as it is prone to the whim of fashion and does not provide a secure base for the wool industry's production. In consequence of these decisions, the wool industry was precluded from enjoying any of the "bandwagoning" effects from the natural growth of the man-made fibres market. The recent decision to "go blended" is unlikely to provide the same financial benefits to the wool industry as in the period 1960-69 because there are indications that the synthetic fibres market has stabilised [9, p. 8].

More recently a suggestion [23] that the International Wool Secretariat should consider a more flexible approach towards promoting blends and that the wool industry might benefit from additional promotion of blends brought a sharp reaction and a note of caution from two members of the International Wool Secretariat [14], [25]. In contrast, the Manager of the Economics Division, Australian Wool Corporation, recently noted a trend in consumer preference towards blends and said:

> Current evidence suggests an increasing trend towards blends. In the past the net effect has been to wool's apparent disadvantage, although it could be argued that wool would have been worse off had blending with synthetics proved impossible. It is not possible to foresee any change to this situation in the future without some strong marketing to ensure wool's share of the blend is at a maximum consistent with consumer requirements [27, p. 25].

There are many aspects of the promotion of fibres including wool which would benefit from systematic analysis. Unfortunately we can only cover a limited number of aspects here. Some of the topics requiring further analysis are mentioned in the concluding Chapter. We hope to explore some of these in the future.

In the remainder of this Chapter we shall consider (i) the relevance for the wool industry of the Dorfman-Steiner and Nerlove-Waugh theorems, (ii) the question of altering consumers' preferences for blends, (iii) the problem of maximising wool sales when the wool industry's promotional budget can be distributed between effort on promoting pure wool and effort on promoting wool blends, (iv) the extent to which competition between manufacturers and processors of man-made fibres can be relied upon in promoting wool blends and (v) the scope for the wool industry to co-operate with manufacturers and processors of man-made fibres in promoting wool blends.

9.3 RELEVANCE OF DORFMAN-STEINER AND NERLOVE-WAUGH THEOREMS TO WOOL PROMOTION

The Dorfman-Steiner theorem asserts that under monopoly conditions, a firm's optimal ratio of advertising expenditure to sales revenue (that is, its profit-maximising degree of advertising intensity) is equal to the ratio of the advertising elasticity of demand for the firm's product to the (absolute) price elasticity of demand for its product [8]. Thus, the more elastic is the demand for a firm's product, the lower, other things equal, is its optimal advertising intensity. The theorem also implies that if a firm's advertising elasticity of demand is constant, the optimal intensity of advertising by the firm rises linearly with the Lerner measure of its degree of monopoly. The Lerner measure of the degree of monopoly is equal to the difference between the price of a product and its marginal cost of production divided by its price.

Apart from any other limitation which it has (for example, it does not take account of the cumulative effect of advertising) the Dorfman-Steiner theorem is not applicable to co-operative or collective promotion of the products of most agricultural industries. This is because most rural industries do not operate as a tight profit-maximising cartel and in many instances, as in the wool industry, the supplies of producers are not controlled. This suggests that theorems developed by Nerlove and Waugh [13] to cover co-operative advertising when supply is not controlled are likely to be more applicable to the wool industry. Nerlove and Waugh consider the optimal advertising intensity for an industry consisting of a large number of firms which aims to maximise its quasi-rent, when supply and other market conditions are not controlled.

They broadly conclude that an industry's optimal ratio of advertising expenditure to sales rises (i) with the inelasticity of the supply of its product, (ii) with the elasticity of demand for its product with respect to advertising and (iii) with falls in the own price elasticity of demand for its product, as each one of these factors varies and other things are held constant.

Relating this theorem to developments in the wool industry, it might be held that as a result of the growth of substitutes for wool the own price elasticity of demand for wool has increased [7]. Hence, on the basis of the Nerlove-Waugh theorem, and supposing other things unaltered, this suggests that the *optimal* ratio of advertising to sales revenue in the wool industry should be declining. But in the wool industry and in the fibre market generally other things have not remained unaltered. New qualities have been imparted to some woollen garments, e.g., machine washability, permanent pleating and consumers are not always aware of the quality of wool blends. In consequence, the possibilities for promoting wool and wool blends may have increased. The elasticity of quantity sold with respect to advertising expenditure may have risen.

But, apart from this, the theoretical basis of the Nerlove-Waugh theorem can be questioned for it assumes that advertising horizontally displaces a Marshallian demand curve, a curve for which the quantity of the product is measured on the X-axis. By contrast if vertical displacement of the demand curve (by a constant) occurs as a result of increased advertising, different conclusions from those of Nerlove and Waugh follow. For example, increases in the own price elasticity of demand for a product raise the industry's optimal *level* of advertising expenditure whereas this level falls in the Nerlove-Waugh case.

This point is important because the Nerlove-Waugh theorem is being used for policy advice on the mistaken view that the profitability of promoting an industry's product *necessarily* falls, other things equal, as the demand for the industry's product becomes more elastic. The Industries Assistance Commission in its *Draft Report on Financing Promotion of Rural Products* [10] considers the theorem to be important for policy purposes. It is, therefore, worthwhile illustrating the dangers posed by the theorem. This can be done by reference to Fig. 9.7.

In Fig. 9.7 market situations for an industry possessing normal supply and demand curves are illustrated and the industry is assumed to be initially in equilibrium at *e*. Assume that a promotion campaign is under consideration for the competitive industry the effect of which is expected to be to shift the industry demand curve upwards. The effect of the campaign on the industry's quasi-rents will depend, as Nerlove and Waugh suggest, on the elasticity of demand for the industry's product.

Assume that as measured at point *e* the industry's demand curve is the relatively inelastic curve D_1 and that the industry's promotion campaign is expected to move this demand curve up to D_2 so that it is parallel to D_1. The campaign will then increase quasi-rents in the industry by an amount equivalent to the area of quadrilateral *aegc*.

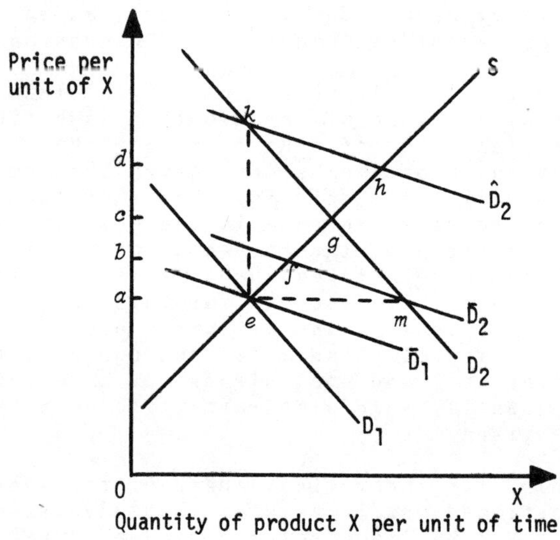

Fig. 9.7

This gain can be compared with the gains for an "equivalent" shift in the demand curve when the initial demand curve is more elastic as measured at *e*. Let the initial (more) elastic demand curve be indicated by $\bar{D}_1$. The result depends upon how the equivalent shift is defined. If an equivalent horizontal shift, equivalent to the distance *em*, is considered and assumed to be achieved with an unchanged level of advertising expenditure, the demand curve moves up to $\bar{D}_2$ and quasi-rents in the industry rise by an amount equivalent to the area of quadrilateral *aefb* an area less than in the more inelastic case in which the gains are equivalent to the area of the quadrilateral *aegc*. This is the type of shift which Nerlove and Waugh consider and in this case advertising relative to existing sales becomes less profitable as demand becomes more elastic. On the other hand, if the equivalent shift to that occurring in the inelastic case is taken to be the vertical distance *ek*, the opposite conclusion follows. In Fig. 9.7 the industry demand curve is shown as shifting from $\bar{D}_1$ to $\hat{D}_2$ and quasi-rents are increased by an amount equivalent to the area of quadrilateral *aehd*. Promotion is much more profitable in the elastic case than in the inelastic case for which quasi-rents rise by the equivalent of the area of the quadrilateral *aegc*. Under these conditions, promotion becomes more profitable and optimal advertising intensities rise as the demand for the industry's product becomes more elastic. This result accords with Donaldson's apparent view [7] that product promotion may be more profitable for a rural industry facing an elastic demand for its product.

Clearly, the result does depend upon the nature of the shift in the demand curve, and so generalisations based upon the Nerlove-Waugh theorem are liable to be misleading. If the elasticity of demand for wool has declined as some writers suggest that it has as a result of increased competition from synthetics, then depending upon the circumstances this can provide a case for higher advertising intensities in the wool industry rather than smaller intensities as the Nerlove-Waugh advertising theorem suggests. Insufficient empirical evidence is available at present to decide this issue, but we must not be misled by theory.

However, it should be pointed out that Professor Schrimper, North Carolina State University, has recently come forward to defend the Nerlove-Waugh theorem against the possibility presented above [19]. Nevertheless upon close examination his defence appears to be inadequate [26].

The above points and those made by Parish in an earlier paper [17] suggest that the Nerlove-Waugh theorem cannot and should not be mechanically applied to decisions about promotion in the wool industry and other rural industries. Wool blends open up complicated promotion options not covered by Nerlove and Waugh in their single product model.

9.4 MAXIMISING WOOL SALES BY PROMOTING BLENDS

There are undoubtedly circumstances in which the development of a preference for wool blends by consumers is in the interest of the wool industry, and circumstances in which it pays the wool industry to promote wool blends. This can be illustrated.

Assume that the wool industry aims to maximise the quantity of wool, W, which it sells for a given outlay on promotion (the available supply of wool being nominally assumed to be in perfectly elastic supply). In doing this, the industry has an option of allocating promotional outlay between $i = 1, \ldots, n$ blends each of which contains the fraction w_i of wool. One of these "blends" may correspond to pure wool ($w = 1$).

Stated mathematically the problem for the industry is to maximise

$$W = \sum_{i=1}^{n} w_i y_i \text{ (where } 0 < w_i < 1) \tag{9.1}$$

$$F(y_1, y_2, \ldots, y_n) = 0 \tag{9.2}$$

where y_i is the quantity sold of the ith blend and F represents the promotion frontier for blends given the "fixed" total promotional outlay of the wool industry. Assuming F to be concave, to be suitably differentiable and ruling out corner solutions, total wool sales, W, will be maximised when the promotional budget is allocated between blends to ensure that

$$w_i + \lambda F_i = 0, \ (i = 1, \ldots\ldots, n) \tag{9.3}$$

$$F = 0, \tag{9.4}$$

where λ is a Lagrange multiplier. Taking any blend, say the nth, the optimal allocation occurs when

$$\frac{F_i}{F_n} = \frac{w_i}{w_n} \ (i = 1, \ldots\ldots, n - 1). \tag{9.5}$$

The left-hand side of this expression represents the marginal trade-off between sales of the ith and the nth blend, given the promotion constraint, and the right-hand side represents the relative wool content of the two blends. The condition implies that the sales trade-off between any two blends must be equal to their relative wool content if promotional outlay is to be optimally allocated. The marginal equilities in (9.5) hold for all those blends it is worthwhile for the wool industry to promote, and the result implies that it can be optimal to jointly promote a range of pure wool and selected blends.

Equation (9.5) also implies that in equilibrium

$$\frac{F_i}{w_i} = \frac{F_n}{w_n} \ (i = 1, \ldots\ldots, n - 1) \tag{9.6}$$

the marginal wool sales from increasing the sales of blend i is equal to the marginal wool sales forgone as a result of the sales forgone of blend n, i and n being any two different blends.

The necessity of this condition being satisfied, given the above conditions, is easily illustrated if we assume two blends, 1 and 2. Let θ_1 represent the increase in the sales of blend one, y_1, which can be achieved by altering promotion in favour of blend one so that sales of blend two decline by one unit. Then, wool sales increase, remain constant or decline for increased promotion of blend one accordingly as

$$w_1\theta_1 \gtreqless w_2 \text{ or as } \theta_1 \gtreqless \frac{w_2}{w_1} \tag{9.7}$$

For instance, let us suppose that blend one contains 0.5 units wool and blend two contains 0.8 units of wool. Then a transfer of promotion from blend two to blend one requires that more than 1.6 *additional* units of blend one be sold for each unit of blend two *forgone* if total wool sales are to rise. However, when both sides of expression (9.7) reach equality, it is impossible to increase wool sales by reapportioning the promotion budget between blends.

It is not difficult to illustrate these conclusions diagrammatically for the case of two alternative blends (one of which may be the "blend" pure wool). In this case, total wool sales by weight are given by

$$W = w_1 y_1 + w_2 y_2 \tag{9.8}$$

which is to be maximised subject to the sales possibility frontier

$$F(y_1, y_2) = 0. \tag{9.9}$$

If the sales possibility frontier for a given promotional budget is concave and differentiable (and corner solutions are excluded), this maximum occurs for

$$\frac{dy_2}{dy_1} = -\frac{w_1}{w_2}. \tag{9.10}$$

In Fig. 9.8, the curve *ABC* represents a possible form of the sales trade-off frontier, *F*, between the blends. The lines marked W_1, W_2, W_3 and W_4 are iso-wool quantity lines. All combinations of the sales of blends along the line W_1 result in the same volume of wool sales W_1. This line is obtained by rearranging

$$W_1 = w_1 y_1 + w_2 y_2 \tag{9.11}$$

to give

$$y_2 = \frac{W_1}{w_2} - \frac{w_1}{w_2} y_1 \tag{9.12}$$

which has a slope of $-\frac{w_1}{w_2}$. The other iso-wool quantity lines may be obtained in the same manner and all have a slope of $-\frac{w_1}{w_2}$, equal to ratio of the proportionate wool content of the blends.

In the case illustrated in Fig. 9.8 the optimal attainable sale of wool blends corresponds to the combination at *B*. Any other attainable combination is on a lower iso-wool quantity line. At

B the slope of the sales trade-off frontier, $\frac{dy_2}{dy_1}$, equals the slope of an iso-wool quantity line, $-\frac{w_1}{w_2}$ and thus the condition stated in (9.10) is satisfied.

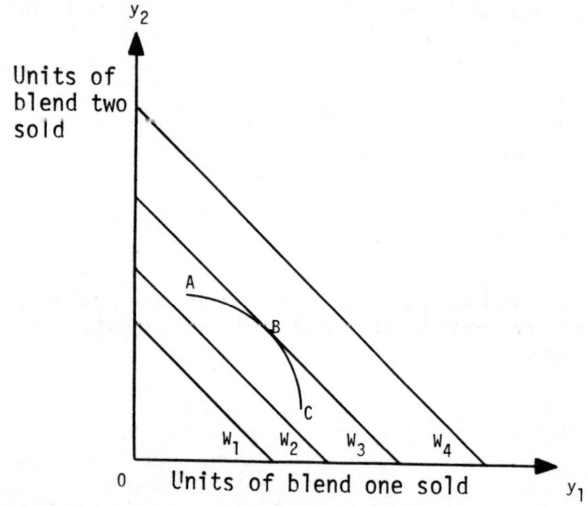

Fig. 9.8

As a particular example, consider the alternative of promoting pure wool ($w_1 = 1$) or a blend ($0 < w_2 < 1$). As long as the increase in the sales of the blended units relative to the pure wool units forgone exceeds the reciprocal of the wool content of the blended units (that is, $\frac{1}{w_2}$), wool sales are increased by allocating more funds to promoting the blend. If for instance $w_2 = 0.5$, the expansion in the sale of blended units must exceed twice any loss in the sale of pure wool units, if wool sales are to increase as a result of the trade-off.

The above analysis rules out corner-point solutions. But sometimes it does not pay the wool industry to promote some blends at all. In terms of Fig. 9.8 the slope of the relevant curves may be such that the optimum is achieved at end-points A or C of the sales trade-off curve. Nevertheless, if maximum wool sales are required and if F is concave and differentiable, then from the Kuhn-Tucker theorem, the marginal equalities stated in (9.5) must hold for all those blends which it is worthwhile for the wool industry to promote.

In the event that the sales possibility frontier is strictly convex rather than concave, (i.e. sales of any and every blend increase at a growing rate with increased promotion of the blend), the optimal strategy for the wool industry is to promote one selected blend which might be, but need not be, pure wool. Looking at the overall picture, any promotional body aiming to maximise the quasi-rents of woolgrowers must, in relation to its expenditure on promotion, push the demand curve for wool as far to the right as is possible. Its promotional strategy must ensure that relative to its expenditure on promotion, the demand for wool is maximised at every price level. As the above theory indicates, this may require the body to promote selected blends. But other actions may also be called for. For instance, the promotional body may need to inform consumers and textile producers of new developments in the treatment of wool. Given that promotion is efficient in the sense that it maximises demand for wool for a given outlay on promotion, or minimises the promotional cost of achieving a specific level of demand for wool, the industry's quasi-rents are maximised by expanding expenditure on promotion until the additional costs involved are just equal to the marginal increase in quasi-rents which are achieved. In a purely competitive world, such as that envisaged in the Nerlove-Waugh model [16], it is theoretically easy to determine the optimal level of promotional expenditure for an industry. But the purely competitive assumption may be inapplicable to the wool industry.

Other fibres are not in pure competition with wool. For instance, this is the case for man-made fibres because the man-made fibre industry is oligopolistic in most countries and dominated by a few multinationals. Promotional effort in the wool industry may be matched and cancelled out to some extent by countervailing promotion by manufacturers of man-made fibres. This needs to be taken into account in determining an optimal promotion strategy for the wool industry. Countervailing expenditure on promotion tends to lower the optimal level of promotion for the wool industry.

9.5 THE PROMOTION OF BLENDS BY COMPETITION BETWEEN MANUFACTURERS OF MAN-MADE FIBRES AND TEXTILE PRODUCERS

We shall argue that the wool industry is unlikely to maximise wool sales and its returns if it relies passively on man-made fibre and textile producers to promote wool blends. Unfortunately, some members of the wool industry believe that it is optimal to leave the promotion of blends entirely to manufacturers of man-made fibres.

Manufacturers of man-made fibres may promote wool blends. If consumers prefer wool blends to pure synthetics and if commodities made from these blends can be marketed at a competitive price, free competition should, *if present*, ensure that wool blends are produced and promoted by textile producers, including those producers with interests in the sale of synthetic fibres. Every maker of man-made fibres has an incentive to increase his market share of man-made fibres and his profit by satisfying consumer demands. This can lead each to promote blends as a competitive strategy. As a result, makers of man-made fibres may as a whole experience reduced sales of synthetics and lower profits.

Makers of man-made fibres may face a dilemma of the prisoner's type [15, p. 95], which is readily illustrated by a duopoly example. Suppose two makers of man-made fibres each of which has the alternative of promoting pure synthetic yarn or a blended yarn containing only some of his synthetic fibre. A producer of man-made fibres may promote a blend by providing suitable advertising subsidies to textile processors who use or make the blend, and by providing these processors with technical advice, samples and so on. Where the synthetic fibre producer owns textile processing plants or firms, as is common in the United Kingdom, a decision can be taken for these plants or firms to produce and promote the blend.

Imagine that there are two makers of man-made fibres and that the pattern of their payoffs for alternative strategies are as indicated in Table 9.1. The arithmetical figures given for the payoffs are purely for illustrative purposes and the mathematical structure of the problem can be generalised [15] to show circumstances in which action on the basis of individual self-interest conflicts with the collective interest of a group. The entry (70,20) for example indicates that if maker 1 promotes a blended yarn and maker 2 promotes only a pure synthetic yarn, maker 1's payoff is 70 and maker 2's payoff is 20. Whatever strategy is adopted by his opponent, it always pays the other competitor, acting in his own self-interest, to promote a blended yarn. Consequently each is driven by his own self-interest to promote a blended yarn and they do not maximise their collective gains. By promoting a blended yarn each producer of synthetics obtains 40 whereas by promoting only a pure synthetic yarn each could have obtained 50. While the competitive self-interest of the duopolists may lead each to promote a blended yarn, the duopolists have an incentive to collude and promote only pure synthetic yarn. In the absence of competition, one would expect the promotion of pure synthetic yarn to be the joint strategy of the producers, to the detriment of wool. The outcome depends to some extent on the strength of competition between producers of synthetics.

Now that a number of important patents on man-made fibres has lapsed the number of makers of man-made fibres has risen and trade barriers have been reduced, e.g., for countries within the E.E.C., competition between makers of man-made fibres has possibly increased. Hence, the production and promotion of blended yarns may increase. The wool industry may find that it is no longer faced by a strong coalition of manufacturers of man-made fibres as might have been so in the past. Competition, however, does not ensure production of blends and their promotion unless consumers prefer blends in comparison to pure synthetics. When the production of synthetic fibres is competitive and consumers do not display a marked preference for blends, no producer of synthetic fibres has an incentive to alter that preference. But if preferences are altered, say by the wool industry convincing customers of the advantage of adding wool to synthetic fibres, competition between makers of man-made fibres and textile producers will, if present, lead to their production and promotion of blended products. Secondary promotion of wool blends by the wool industry can lead to primary promotion of these by makers of man-made fibres.

TABLE 9.1

MAKER 1'S STRATEGIES	MAKER 2'S STRATEGIES Promote pure synthetic	 Promote blended yarn
Promote pure synthetic	(50,50)	(20,70)
Promote blended yarn	(70,20)	(40,40)

When a new synthetic fibre is being introduced consumer preferences may be such that acceptability of the new fibre is only gained by using it in mixtures with other fibres. Once acceptability of the new fibre is established the proportion of natural fibre used in blends may be reduced by the maker of the new synthetic. Generally the wool industry has no interest in establishing new fibres, except in cases where they are complementary to wool. But once a fibre becomes established the industry has an interest in the proportion of wool used in any blend containing the fibre and may be able to influence this by convincing consumers or textile processors of the advantage of maintaining the wool content of any blend. The industry may be able to promote wool successfully in this way while pointing out that the optimal proportion of wool may vary for different end-uses [12] and that wool-rich blends and pure wool are not appropriate for some end-uses.

Within the above context of collusion or non-collusion between manufacturers of synthetics, there is little scope for the wool industry to co-operate with makers of synthetics in promoting wool blends. Furthermore, it would be a folly for the wool industry to abandon the promotion of blends to the man-made fibre industry because of the potential for changes in the proportions of various fibres in the blends. The consequences of such a move would be under-representation of the advantages of using wool in blends and the likely denial to the wool industry of a partial free-ride. This partial free-ride may be obtained if the wool industry convinces textile processors and the consumers of the advantages of maintaining wool content in a blend and if fibre producers compete to satisfy the preferences of buyers. However, the possible reactions of manufacturers of synthetics must be taken into account. Even if manufacturers form only a loose coalition, they may react to a rise in the wool industry's expenditure on the promotion of selected wool blends by increasing their own expenditure on the promotion of synthetics. Nevertheless, after taking account of this possible reaction, the wool industry may gain by expanding its promotion of selected blends.

Yet it needs emphasising that Table 9.1 indicates only one set of possibilities - a set where promotion of blends by makers of man-made fibres forms a stable equilibrium because of competition between them. If we were to reverse the entries in the left-hand top and right-hand bottom positions of this Table, competitive makers of man-made fibres would achieve a stable equilibrium by promoting pure synthetic yarns even though their profit would be less than if both promoted blends.

One cannot say *a priori* whether wool sales are least reduced by makers of man-made fibres promoting blends or by their promoting pure synthetics. The greater the extent to which synthetics are accepted as substitutes for wool or are considered superior to wool, the more damaging to wool sales is likely to be the promotion of pure synthetics compared to blends.

TABLE 9.2

Example of Variation of Payoff Matrix to Alter Equilibrium.

The Wool Industry may for instance be able to Transform Matrix I to Matrix II

MATRIX I

	MAKER 2'S STRATEGIES	
MAKER 1'S STRATEGIES	Promote pure synthetic	Promote blended yarn
Promote pure synthetic	(50,50) [(a)]	(60,40)
Promote blended yarn	(40,60)	(30,30)

MATRIX II

	MAKER 2'S STRATEGIES	
MAKER 1'S STRATEGIES	Promote pure synthetic	Promote blended yarn
Promote pure synthetic	(50,50)	(45,55)
Promote blended yarn	(55,45)	(40,40) [(b)]

Note: (a) Competitive stable equilibrium.
(b) Competitive stable equilibrium.

While the ability of the wool industry to do this should not be exaggerated, it may be possible for the wool industry to alter the pattern of payoffs to makers of man-made fibres (by influencing consumer preferences) so that the equilibrium competitive behaviour of makers of synthetic fibres alters in a way which expands sales of wool. For instance, let us suppose that in this particular example the promotion of blends maximises wool sales but that matrix I in Table 9.2 applies so that the promotion of pure synthetics is the strategy adopted by makers of synthetics. By doing some advertising of blends, the wool industry may be able to transform this matrix to a type like II in Table 9.2. The promotion of blends by *competitive* makers of synthetics *now* forms a new stable equilibrium - one in which makers of synthetics are worse off than previously. In some circumstances, a *small* input by the wool industry can conceivably lead to a *major* shift in market equilibrium once the competitive reactions of makers of synthetic fibres come into play.

In some circumstances of course the promotion of blends by makers of synthetics does not maximise wool sales. It may be possible in these circumstances for the wool industry to modify the behaviour of makers of synthetics if they are promoting blends so that they concentrate on the promotion of synthetics. For instance, imagine that the column headings are interchanged and also the row headings are interchanged in Table 9.2. In this case the modification of matrix I shifts makers of synthetics from promoting blends to promoting pure synthetics. The wool industry might achieve the modification in matrix I by "knocking" blends, that is by convincing enough individuals that blends are an inferior alternative. Whether or not it is to the advantage of the wool industry to support blends or "knock" them is likely to vary from product to product.

9.6 CONCLUSION

It is probably profitable for the wool industry to promote a range of selected wool blends, limiting its promotional support in certain circumstances as with Woolblendmark, despite the existence of a "trickle-down" effect from the promotion of pure wool. Furthermore, it would seem to be unwise for the wool industry to rely solely on manufacturers of man-made fibres and textile producers to promote wool blends. The success which the wool industry is likely to achieve by promoting blends will depend partly on the state of competition between manufacturers of man-made fibres and between producers of textiles. As demonstrated in this Chapter, the wool industry may be able to improve its own gains by taking advantage of such competition as exists. While oligopolistic competition from manufacturers of man-made fibres and organisations representing other fibres needs to be reckoned with by the wool industry, it is probably less true now than it used to be that manufacturers of man-made fibres are in collusion. Nevertheless, game-like models continue to be more relevant to analysing the wool industry's promotional strategies than the purely competitive model of advertising developed by Nerlove and Waugh.

REFERENCES

[1] Australian Bureau of Statistics, *Yearbook 1974*, Canberra, 1975.

[2] Australian Bureau of Statistics, *Yearbook 1975-76*, Canberra, 1976.

[3] Australian Bureau of Statistics, *Quarterly Summary of Australian Statistics*, Canberra, March, 1976.

[4] Australian Wool Corporation, *Interim Annual Report, 1975-76*, Melbourne, 1976.

[5] Bureau of Agricultural Economics, *The Fibre Review 1975-76*, Australian Government Publishing Service, Canberra, 1977.

[6] De BOER, A.J., "Rural Product Promotion: Economic Aspects of Promotability, Organisation, and Public Assistance", *Review of Marketing and Agricultural Economics*, 45, 121-145, 1977.

[7] DONALDSON, G.F., "The Financing of Agricultural Research Levies on Farm Produce", *Review of Marketing and Agricultural Economics*, 32, 183-216, 1964.

[8] DORFMAN, R., and STEINER, P.W., "Optimal Advertising and Optimal Quality", *American Economic Review*, 44, 826-836, 1954.

[9] EDWARDS, T.V., and TRICKETT, E.S., "A Systematic Analysis of the Dangers in Australian Primary Production and Strategies for their Future Development", (mimeo), *Second Conference of Economists*, Sydney, August, 1971.

[10] Industries Assistance Commission, *Draft Report on Financing Promotion of Rural Products*, Canberra, June 1976.

[11] International Wool Secretariat, *Die Optimale Mischung Gibt es Nicht*, Dusseldorf, 1975.

[12] International Wool Secretariat, *Promotion in Action*, London, 1975.

[13] JOYCE, T., "Advertising", in A.S.C. Ehrenberg and F.C. Pyatt (eds) *Consumer Behaviour*, Penguin, Harmondsworth, 1971.

[14] LEWIS, N., and BEREZA, S., "The Promotion of Wool-Synthetic Fibre Blends: A Comment", *Review of Marketing and Agricultural Economics*, 45, 95-98, 1977.

[15] LUCE, R.D., and RAIFFA, H., *Games and Decisions*, John Wiley, New York, 1966.

[16] NERLOVE, M., and WAUGH, F.V., "Advertising Without Supply Control: Some Implications of a Study of Advertising of Oranges", *Journal of Farm Economics*, 43, 813-837, 1961.

[17] PARISH, R.M., "Possibilities for Promoting Farm Products", *Australian Journal of Agricultural Economics*, 7, 27-34, 1963.

[18] ROBERTS, R.C., *A Review of Developments in Man-Made Fibres 1960-1971*, Wool Economic Research Report No. 24, Bureau of Agricultural Economics, Canberra, 1973.

[19] SCHRIMPER, R.A., "Comments on Tisdell's Critique of Nerlove-Waugh Theorem Concerning Optimal Advertising", *Review of Marketing and Agricultural Economics*, 45, 186-189, 1977.

[20] STURGESS, I.M., "Marketing Policies for Rural Industry", in D. Douglas (ed.) *National Rural Policy*, University of Sydney Extension Board, Sydney, 1971.

[21] TISDELL, C.A., "Research and Promotional Levies for Australian Rural Products", *Farm Policy*, 14, 63-75, 1974.

[22] TISDELL, C.A., "Promotion and Advertising by Monopolies and Cartels - A Neglected Welfare Aspect", *Research Report or Occasional Paper No. 22*, Department of Economics, University of Newcastle, November 1975.

[23] TISDELL, C.A., "The Promotion of Wool and Synthetic Fibre Blends: Some Alternative Strategies for the Wool Industry", *Review of Marketing and Agricultural Economics*, 44, 101-113, 1976.

[24] TISDELL, C.A., "Is Advertising Expenditure Socially Excessive?", *Bulletin of Economic Research*, 29, 57-60, 1977.

[25] TISDELL, C.A., "A Rejoinder on the Promotion of Wool-Blends", *Review of Marketing and Agricultural Economics*, 45, 99-100, 1977.

[26] TISDELL, C.A., "Rejoinder on the Nerlove-Waugh Advertising Theorem", *Review of Marketing and Agricultural Economics*, 45, 190-191, 1977.

[27] WARD, L.E., "The Future of Wool as a Textile Fibre", *Farm Policy*, 17, 24-29, 1977.

[28] Wool Marketing Committee of Enquiry, *Report*, Canberra, February 1962.

CHAPTER 10

Concluding Comments

The purpose of this monograph has been to consider economic aspects of fibre markets with particular relevance to wool. Markets for the various types of fibres undoubtedly exhibit a great deal of interdependence. After exploring the nature of fibre markets and countervailing competition in the fibre industry, a major part of the study was concerned with the possible impact of the growth in the market share of man-made fibre industry upon fluctuations in the prices of natural fibres. The study suggests that important interrelationships do exist between fluctuations in the prices of natural fibres and capacity utilisation and market share of man-made fibres.

The way in which textile producers adapt to fluctuations in prices of natural fibres, especially wool, was explored and attempt was made, using survey data, to assess the value which textile producers place on stability of fibre prices. In general, the further the textile manufacturer tended to be along the textile production chain the more likely was it that he valued price stability or certainty. We also correlated textile manufacturers attitudes to price stability to other factors such as their ability to alter fibre mixes and found significant results. For instance, textile manufacturers who are more easily able to vary their mixtures of fibres tend to be more frequently in favour of price stability of fibres. This result can be expected on theoretical grounds, for example, the relationship between profits, flexibility and price uncertainty [13], [14]. Furthermore, as a part of our consideration of the effects of instability of wool prices on the decisions of textile manufacturers, the holding of wool stocks by textile producers was related to expectations of these producers about wool prices, and attitudes of textile producers towards the wool futures market were examined using a survey of Australian textile producers.

While an analysis of aspects of wool promotion formed a minor part of the study, the analysis enabled us to pinpoint some dangers involved in using Nerlove's theorem to determine the optimal intensity of advertising, and to explore some of the consequences for wool sales of promoting fibre mixtures, particularly in the context cf interfibre rivalry with synthetic fibres. However, there are many other aspects of wool and fibre promotion which require urgent study from an economic standpoint. The following comment in the "Green Paper", *Rural Policy in Australia,* still seems relevant even though the Industries Assistance Commission (I.A.C.) has now reported [9] on wool promotion:

> It would clearly be desirable for there to be more public evaluations made of the nature and effectiveness of wool promotion in all its stages. Undoubtedly, technological advances for wool need promoting at all stages in the fragmented and complex processing and marketing chain, to ensure best use is made of them and that consumers are made aware of improvements in final products. At present the case for wool promotion however, rests largely on the considerable resources spent on man-made fibre promotion which the wool industry needs to counter [8, Sec. 6.224].

The I.A.C. in its recent enquiry into the promotion of Australian rural products [9], attempted to assess returns to wool promotion. The Commission received evidence from the International Wool Secretariat (I.W.S.) indicating a marginal response of six kilograms of clean wool per $A (1975 prices) spent on wool promotion, and the I.A.C. suggested on the basis of econometric evidence that the break-even response to wool promotion from Australia's viewpoint was 0.5 to 1.2 kgs. per $A. The I.A.C. expressed the view that on the basis of all the evidence available to it, Australia would probably gain from additional promotion of wool. Detailed econometric study of wool sweater promotion led the I.A.C. to conclude that the marginal response of clean wool sales per $A spent on wool sweater promotion in 1974 was 3.5 kg. The studies of the I.A.C. about returns to wool promotion need to be extended to cover returns to promoting wool in more end-uses.

Attention also needs to be given to the returns which the wool industry can obtain by promoting products in co-operation with textile producers. At the present time the I.W.S. partly finances the promotion campaigns of some textile producers and distributors of wool products and the benefits to be achieved from doing this require economic analysis.

It may also be true that systematic consequences for wool promotion and costs associated with the handling of wool flow from the fact that it is becoming *relatively* less important in most end-uses. For example, is there a possibility that a wide-range of wool garments may in the end be only offered by specialty retailers and in very large department stores?

Policies in relation to wool research and fibre research generally also raise many unresolved issues. There appear to be no mechanisms which automatically ensure that research funds are allocated in a socially optimal way [15], nor even in a way which maximises on industry's benefits from its research effort [7]. In the case of the Australian wool industry, for example, research funds have been geared to a considerable extent to the value of the wool clip and practically all of these funds have been specifically directed to wool research. With the possible decline of the Australian wool industry, one wonders whether some of these funds might not be most usefully employed in exploring alternative economic opportunities for woolgrowers in marginal areas. This is not to deny the need and value of further research specifically on wool but to suggest that even from a general welfare point of view of those in the wool industry a partial single product focus in research may not be optimal. In relation to this matter, the "Green Report" stated:

> There is a logic in the argument that the industry funds [for R & D] provide an inflexible system in many respects, although there are corresponding benefits which tend to be overlooked. The inflexibility of the system could be partly overcome by enabling and encouraging these funds, as some already do, to incur expenditure on products or activities such as pasture improvement likely to improve the productivity of the land and other resources of their contributors, rather than restricting research to the commodity on which levies were collected [8, Sec. 5.144].

Since the "Green Report" the financing of rural research in Australia has been investigated by the I.A.C. The recommendations in the I.A.C.'s report on research [10] would, if adopted, strengthen the control of woolgrowers over the use of funds which they contribute for wool research through compulsory levies on their clip. At present the Australian Government subsidises this research and while the Commission has not recommended a reduction in governmental support, it has recommended changes in the method by which the government's contribution to wool research is allocated.

The level of demand for wool (the demand schedule) which can be influenced by promotion and research, is only a partial determinant of the share of the fibre market enjoyed by wool. In the long-term, costs and alternative profit opportunities for woolgrowers exert an influence on supplies for wool and the industry needs to predict trends in these to plan for the future.

In Australia there has been a long-term tendency for the terms of trade to move against wool - prices received by farmers for wool relative to prices paid by them for inputs have tended to fall (see Fig. 10.1). Partly as a result of the declining terms of trade for wool, the number of sheep holdings in Australia has been declining more rapidly than the general decline in rural holdings (between 1959-60 and 1970-71 the former declined by 7.97 per cent and the latter by 1.09 per cent), and sheep numbers and the volume of wool production have been declining in recent years (see Fig. 10.2). Whether or not this decline is temporary remains to be seen but the possibility of a long-term slow decline in the Australian wool industry cannot be dismissed even though in the foreseeable future Australian wool production should continue to be substantial and wool should remain an important minority fibre in the fibre market.

In assessing the likely supply of Australian wool one needs to take into account the profitability of alternative industries such as beef, wheat and other grains. While beef prices are currently depressed they are expected to improve in the long-term and demand for grains is anticipated to grow. "It appears that both in the beef and wheat areas that the long-term trend will be against wool, although not to a dramatic extent" [16, p. 29].

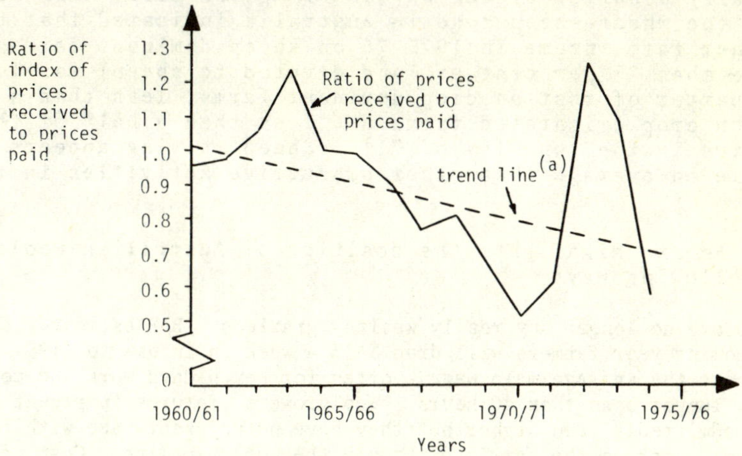

Fig. 10.1

Ratio of Index of Prices Received by Farmers for Wool to Index of Prices Paid for Inputs (Base: Average 1960-61 to 1962-63)

Note: (a) Trend Line fitted by method of least squares.

Source: Based on statistics in *Quarterly Review of Agricultural Economics* [3] (various issues).

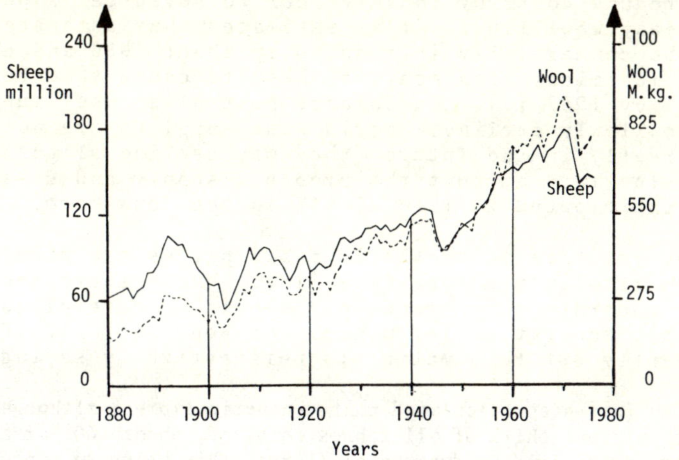

Fig. 10.2

Sheep Numbers and Wool Production[a] Australia 1880-1977

Note: (a) Wool on greasy basis.

Source: Based on Plate 53 in *Official Yearbook of Australia, 1975-76* [1], Canberra, 1975.

Incidentally a survey by the Bureau of Agricultural Economics of farms in the wheat-sheep zone in Australia indicated that the *average* net farm income in 1975-76 on sheep dominant farms (farms with more than 75 per cent of land devoted to sheep) was less than a quarter of that on crop dominant farms, less than a third of that on crop orientated farms and less than a half of that on mixed farms in the zone [4, p. 71]. Sheep raising appears less profitable on average than other productive activities in the zone.

The *Wool Record* highlights the position of Australian woolgrowers in the following way:

> There are no longer any really wealthy graziers. BAE estimates that in the coming year farmers will drop $A13 a week in income to $199 - just $9 above the average male wage - often for seven days work and certainly for a longer span than 40 hours. Woolgrowers' returns in recent years have admittedly been higher but they have no way kept pace with cost increases, both on the farm and through the wool pipeline. Cost of shearing and getting wool to store has risen from 18.21 cents a kg. greasy to 55.95 cents in six years. Total cost of getting wool from the sheep's back to mills has risen by 143% in the same period compared with a 76% rise in the cost of living [5, p. 65].

While new developments are occurring which should reduce the cost of wool (for example, chemical shearing and automated shearing are likely in the future) a dramatic and sudden reduction in costs of production is not expected on this account. For instance new shearing techniques when they become operational are likely to reduce costs by roughly four to seven cents per kilogram of greasy wool [16]. It is estimated that automated shearing might be commercially introduced by about 1980 and chemical harvesting (chemical treatment of sheep to cause them to shed their wool) by 1982 [6]. On balance Australian wool supplies are expected to slowly decline. World wool supplies are not expected to alter greatly in the future (they may decline slightly). Wool prices (taking into account the probable supply and demand position) are anticipated to rise slowly in the long-term.

Wool prices relative to synthetic fibre prices are predicted to continue to rise in the long-term [17]. While increased oil prices may moderate this trend, one should be careful not to exaggerate this moderating influence. Mr Bonus of C.I.R.F.S., Paris, recently put this matter in perspective by saying:

> It has in fact been calculated that synthetic fibres, although they represent only one-third of all fibres combined, absorb 60% of all the energy used in world production of fibres, this being so since the production of 1 tonne of man-made fibres requires 5 tonnes of oil, while 1 tonne of natural fibres only requires 1 tonne of oil. Yet, although this figure of 5 tonnes may seem very impressive, man-made fibres production actually hardly represents 2% of total consumption of refined oil. Du Pont de Nemours, for their part, estimated in 1974 that the total yearly needs for synthetic fibres of a 3-person American family could be met by a quantity of energy equivalent to the contents of their car's petrol tank. Besides, the value of the contents of this petrol

> tank is quite low compared to the value of the synthetic fibres potentially obtainable from this quantity of petrol. Energy, even if it were to become much dearer, is not an insuperable handicap for the development of man-made fibres. It is however a major problem for fibre producers since the general economic situation does not always allow them to compensate for the higher price of raw materials through their selling price [2, p. 45].

Another influence on relative prices will be the speed and extent to which the synthetic fibre industry can reduce excess capacity and rationalise. This appears to be gathering momentum in the E.E.C. and Japan. In Japan the government has agreed to mergers which will reduce the number of major synthetic fibre producers from eight to four and capacity by 28 per cent [12]. It is reported that in the E.E.C. [11] synthetic fibre producers are forming a cartel under special provisions which made these legal in emergency conditions. Arrangements within the E.E.C. are expected to involve market sharing and import protection agreements and provisions. Consequently the prices of synthetic fibres may recover more quickly than was originally predicted and reduce the predicted differential between the price of wool and the prices of synthetics.

The future of wool cannot be assessed in isolation. Sensible planning for the wool industry can only proceed, as this monograph illustrates, in an interdependent context - a context of interdependence between fibres, of competition for resources used in wool production and of trading interrelationships between nations. The wool industry has been quick to recognise this, quicker than the cotton industry and has benefited as a result. It would be unfortunate for the wool industry if the Australian industry or that of any other major wool producer should as a result of the recent recession and the possibility of a renewed one, lose sight of this interdependence and withdraw into isolation to save funds.

REFERENCES

[1] Australian Bureau of Statistics, *Yearbook of Australia 1975-76*, Australian Government Publishing Service, Canberra, 1977.

[2] BONUS, J., "Man-Made Fibres and Natural Fibres: Competition in End-Uses in Western Europe", a paper presented at a conference on *The Future of Natural Fibres*, Shirley Institute, Manchester, November 1977.

[3] Bureau of Agricultural Economics, *Quarterly Review of Agricultural Economics*, Australian Government Publishing Service, Canberra (various issues).

[4] Bureau of Agricultural Economics, *Natural Agricultural Outlook Conference 1978*, Australian Government Publishing Service, Canberra, 1978.

[5] Business Press Ltd., "Anxious Moments Ahead for Australian Woolman", *Wool Record*, June, 65, 1977.

[6] Business Press Ltd., "Little Hope Yet of Cutting Wool Harvesting Costs", *Wool Record*, June, 31. 1977.

[7] DUNCAN, R.C., "The Gains from Pasture Research in Australia: An Economic Analysis of Research in the C.S.I.R.O. Division of Plant Industry", unpublished Ph.D. thesis, Australian National University, February 1972.

[8] HARRIS, S., et al., *The Principles of Rural Policy in Australia: A Discussion Paper*, Australian Government Publishing Service, Canberra, 1974.

[9] Industries Assistance Commission, *Draft Report on Financing Promotion of Rural Products*, Australian Government Publishing Service, Canberra, 1976.

[10] Industries Assistance Commission, *Financing Rural Research*, (June 25, 1976) Australian Government Publishing Service, Canberra, 1976.

[11] Newsweek Inc., "Fibre Producers Organize Cartels", *Newsweek*, May 15, 44, 1978.

[12] Textile Business Press, "Japanese Have Problems Too", *Textile Month*, February, 23, 1978.

[13] TISDELL, C.A., *The Theory of Price Uncertainty, Production and Profit*, Princeton University Press, Princeton, 1968.

[14] TISDELL, C.A., "Implications of Learning for Economic Planning", *Economics of Planning*, Vol. 10, 1970, pp. 177-192.

[15] TISDELL, C.A., "Research and Development Services", Ch. 8 in K. Tucker (ed.) *Economics of Australian Service Sector*, Croom Helm, London, 1977.

[16] WARD, L.E., "The Future of Wool or a Textile Fibre", *Farm Policy*, 17, 24-29, 1977.

[17] WILCOX, J., "Why Wool?", a paper presented at a conference on *The Future of Natural Fibres*, Shirley Institute, Manchester, November 1977.

Appendices

APPENDIX I : Generalisation of Theorems by Oi and Tisdell on the Effect of Price Fluctuations on Average Profit.

APPENDIX II : Does Price Instability Increase Consumers Welfare as Waugh and Massell Suggest?

These two appendices deal generally with theoretical effects on the price instability of a commodity or commodities on the welfare of producers and consumers. These appendices are relevant to an assessment of the possible impact of price stabilisation schemes for wool or cotton on the incomes of wool and cotton growers and the welfare of buyers of these products. However, they do not deal with the uncertainty aspect of price instability.[1] This aspect can be very important in practice and needs further investigation.

---

[1] For some further discussion on this aspect see:

BLANDFORD, D., and CURRIE, J.M., "Price Uncertainty - The Case for Government Intervention", *Journal of Agricultural Economics*, 26, 37-51, 1975.

SCHMITZ, A., SHALIT, H., and TURNOVSKY, S.J., *Producer Welfare and Price Stability* (mimeo), University of California, Berkeley and Australian National University, Canberra, March 1978.

TISDELL, C.A., *The Theory of Price Uncertainty, Production, and Profit*, Princeton University Press, Princeton, 1968.

TURNOVSKY, S.J., "A Distribution of Welfare Gains from Price Stabilisation: A Survey of Some Theoretical Issues", paper presented at Ford Foundation Conference on *Stabilising World Commodity Markets*, Airlie, Virginia, March 1977.

APPENDIX I

Generalisation of Theorems by Oi and Tisdell on the Effect of Price Fluctuations on Average Profit

AI.1 INTRODUCTION

In 1961, Walter Oi [1] showed that the average profit of single product perfectly competitive firms is increased by instability of their product's price. This is so if firms' marginal costs are not subject to shifts, for instance, as a result of random disturbances, if their supply curves are not perfectly inelastic and if price uncertainty is not a material consideration [2], [3]. In 1970, Tisdell [4] pointed out that a corresponding theorem, subject to similar qualifications, hold for the instability of a factor's price, namely, that a perfectly competitive firm's average profit is greater for a fluctuating price of a factor than for a stable one equal to the average value of this factor's variable price.

The purpose of Appendix I is to combine these results and present a theorem that covers the case of multi-commodity perfectly competitive firms. Appendix I's contribution lies in its statement of a general price-instability theorem for an n-commodity perfectly competitive firm.

Since many firms produce several products the prices of which fluctuate, there is a need to see whether Oi's proposition generalises. In Australia, for instance, farmers frequently produce both wool and beef and other products such as oil seeds, the prices of which are subject to fluctuation. Furthermore, some firms experience variations in the prices of the multiple inputs which they use. For example, manufacturers of stock food find that the prices of grains, lucerne and other inputs, which they use, fluctuate and textile producers experience variations in natural fibres of different grades. Some firms find that the prices of a number of their inputs and a number of their products vary and it is, therefore, worthwhile considering the multi-commodity case. But, before doing this, it may be useful to restate the main findings of Oi and of Tisdell for a single product and for a single input subject to fluctuations in their prices.

AI.2 SINGLE COMMODITY CASE

Oi [1] assumes that the price of a perfectly competitive firm's product (a single-product firm) fluctuates but that the prices of its inputs are stable. Under these conditions, which imply that the firm has perfect knowledge about its prices and costs, the profit which the firm obtains from fluctuating product prices over an interval of time is greater than the firm can obtain if the price of the product is stabilised at the average of its fluctuating level.

The argument can be illustrated by means of Fig. AI.1. In Fig. AI.1 the curve marked MC represents the marginal cost experienced by the competitive firm producing product X and over an interval of time, the price of this product is assumed to be p_1 50 per cent of the time and p_2 in the remainder of the time. When the price of the product is p_1 the firm's profit is equal to the area of the triangle indicated by B. When the price of the product is p_2, the firm's profit is equal to this area plus the area of the hatched and the shaded quadrilaterals plus the area of the triangle indicated by G. Profit on average for each period of time is equal to half of profit when price is p_1 plus half of profit when price is p_2. In Fig. AI.1 this is equivalent to the area of triangle B plus the area of the hatched quadrilateral (equals the shaded quadrilateral) plus half the area of triangle G.

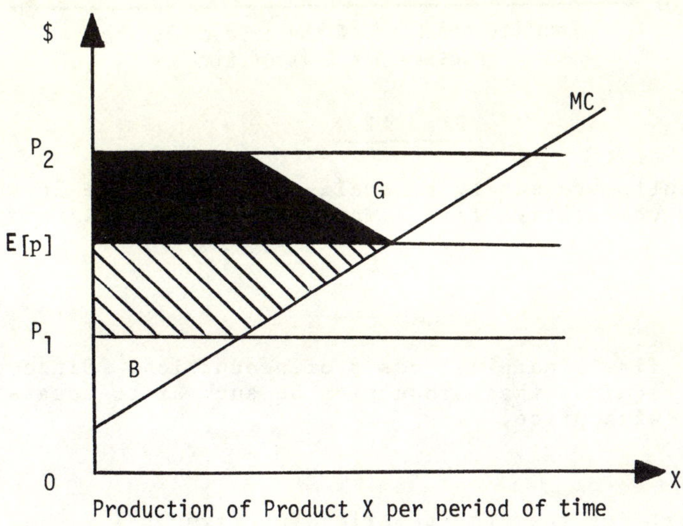

Fig. AI.1

If price is stabilised at $E[p] = 0.5p_1 + 0.5p_2$, the profit of the firm per period is equivalent to the area of triangle B plus the area of the hatched quadrilateral. Thus in the stabilised case the firm's profits per period are lower on average by half the area of triangle G than those when prices are fluctuating. The corresponding industry position to that shown in Fig. AI.1 for the firm could be as illustrated in Fig. AI.2. The fluctuations in the price of the product as shown are arising from variations in the demand for the product.

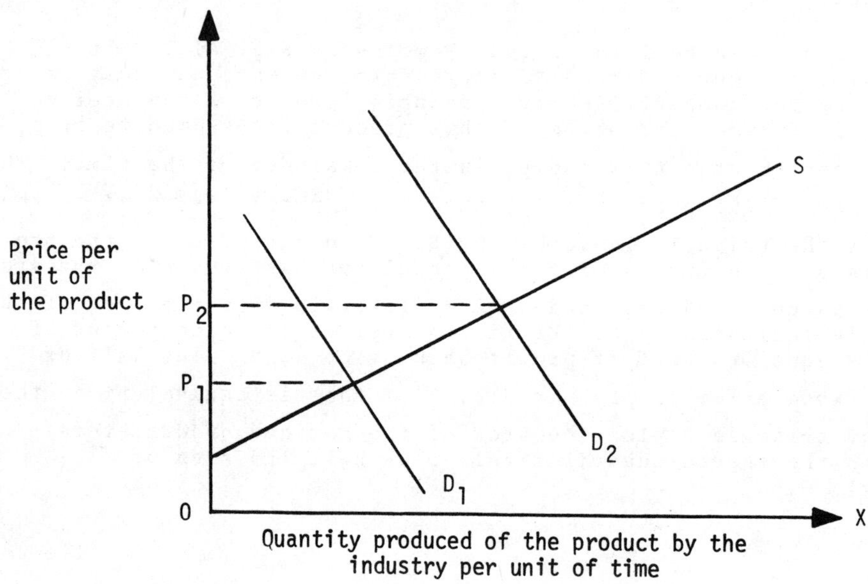

Fig. AI.2

The above results are easily generalised for the case in which the perfectly competitive firm's marginal costs are linear. Let

$$\frac{dC}{dx} = a + bx \qquad \text{(AI.1)}$$

represent the firm's marginal costs of production. Since profit maximisation requires that production be such as to equate the marginal cost with price,

$$p = a + bx \qquad \text{(AI.2)}$$

and, therefore, for the profit-maximising firm

$$x = \frac{p - a}{b}. \qquad \text{(AI.3)}$$

Consequently, as indicated from Fig. AI.3 the formula for the area of a triangle (an appropriate one is hatched in Fig. AI.3) can be used to express the firm's per period profit as a function solely of p, the price of its product. At price p the area of the triangle representing the firm's profit is

$$\Pi = 1/2(p - a)x = 1/2(p - a)\frac{p - a}{b} \qquad \text{(AI.4)}$$

$$= 1/2b(p^2 - 2ap + a^2). \qquad \text{(AI.5)}$$

Thus profit per period on average is

$$E[\Pi] = 1/2b(E[p]^2 + \text{var } p - 2aE[p] + a^2). \qquad \text{(AI.6)}$$

If the average price of the product is held constant but its variance is increased, the profit of the firm per period and on average rises because

$$\frac{\partial E[\Pi]}{\partial \text{ var } p} = 1/2b. \qquad \text{(AI.7)}$$

Profit per period (averaged) is at a minimum if the price of the product is stabilised at the average of its fluctuating prices because as (AI.6) indicates, E[Π] is a rising linear function of the variance of p. Thus, as indicated by Oi, price instability increases the firm's profit under the above conditions.[1]

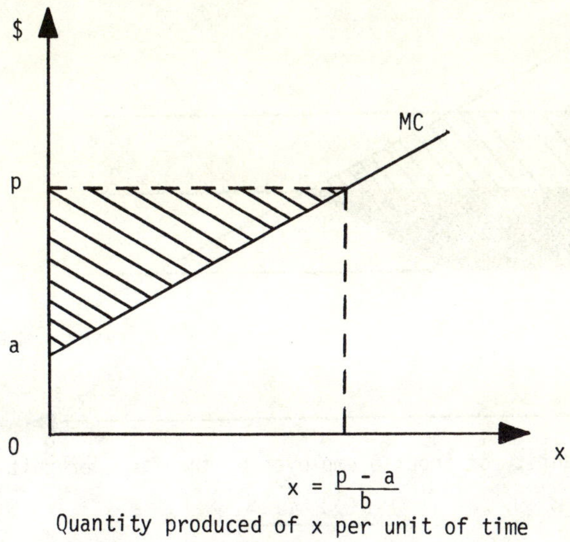

Fig. AI.3

[1] Oi's results apply provided the marginal cost curve of the firm eventually increases [3, pp. 103-105].

An analogous argument applies in the case of a perfectly competitive firm using a single variable input, the price of which is subject to fluctuation. Increased variability of the variable input's price about the same average value raises the firm's profit over an interval of time, all other prices constant. This proposition can be illustrated by means of Fig. AI.4.

In Fig. AI.4 the curve of the value of the marginal product of input Q to a perfectly competitive firm is indicated by the curve marked VMP and it is assumed that the price of input Q is w_1 50 per cent of the time and w_2 in the remaining time. When the price of the input is w_2, the per period profit of the firm is equivalent to the area of triangle B. When the price of the input is w_1 the per period profit of the firm is equivalent to the area of triangle B plus the area of the hatched and the shaded quadrilaterals (which are of equal area) and the area of triangle G. Hence, if the price of factor is w_1 50 per cent of the time and w_2 50 per cent of the time, the firm's profit per period on average is equivalent to the area of triangle B plus the area of hatched quadrilateral plus half of the area of triangle G.

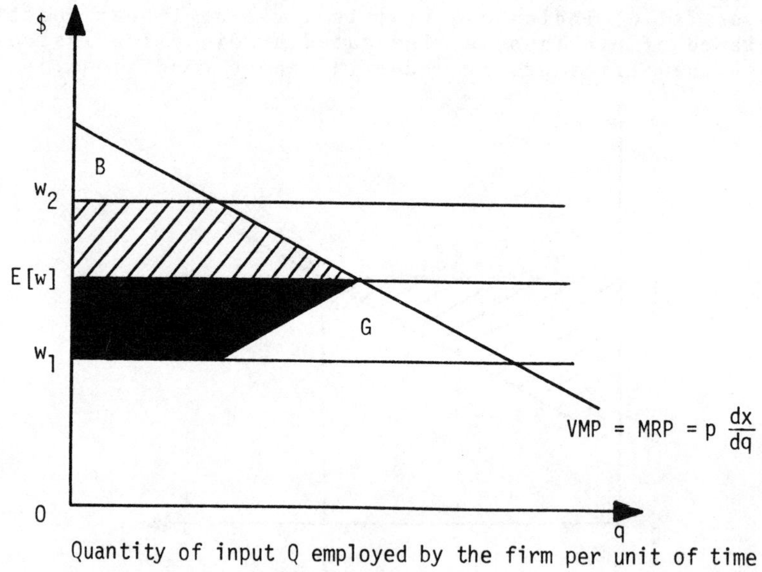

Fig. AI.4

If the price of factor Q is established at E[w], the average of its fluctuating value, the per period profit of the firm is equivalent to the area of triangle B plus the area of the hatched quadrilateral. Comparing this area with that described in the last paragraph, the profit of the firm on average per period is greater when the price of the input fluctuates than when this price is stable. Average profit in the former case exceeds that in the latter case by half the area of triangle G.

This result is easily generalised to cover all cases in which the curve of the firm's value marginal product of its variable input is linear. In the linear case, the firm's profit can be represented by the area in appropriate triangle. Such a triangle is indicated by the hatched area in Fig. AI.5 which indicates the firm's maximum profit when the price of its input is 2. The area of this triangle can be expressed in general terms.

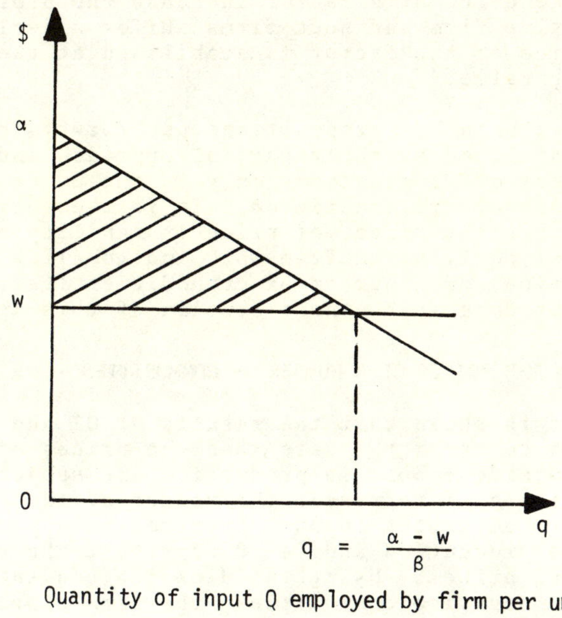

Fig. AI.5

Let $\alpha - \beta q$ represent the firm's value of marginal product from employing Q. Since a profit maximising firm will always employ Q so as to equate Q's price to the value of its marginal product,

$$w = \alpha - \beta q \qquad \text{(AI.8)}$$

and

$$q = \frac{\alpha - w}{\beta}. \qquad \text{(AI.9)}$$

Hence, the firm's maximum profit as a function of w, the area of the appropriate triangle is

$$\Pi = 1/2(\alpha - w)q = 1/2\beta(\alpha^2 - 2\alpha w + w^2). \qquad \text{(AI.10)}$$

Therefore,[2] profit per period on average is

$$E[\Pi] = 1/2\beta(\alpha^2 - 2\alpha E[w]^2 + \text{var } w). \quad \text{(AI.11)}$$

It follows that an increase in the variability of the price of the variable input, Q, raises the profit of the firm, other things equal, because

$$\frac{\partial E[\Pi]}{\partial \text{ var } w} = 1/2\beta > 0 \quad \text{(AI.12)}$$

Fluctuations in the price of a factor increase the profits of a perfectly competitive firm and such firms suffer a decline in profits if the price of the factor is stabilised at the average of its fluctuating value.[3]

It is apparent that both the propositions put forward by Oi and by Tisdell are restricted by their partial approach and by the fact that the effect of fluctuations only in the price of one product or in one input are considered. It is theoretically desirable to consider the effect of price instability on average profit in a general equilibrium framework and to allow for variability in the prices of a number of commodities (factors and products). This is done in the next section of this Appendix.

AI.3 FLUCTUATIONS IN THE PRICES OF A NUMBER OF COMMODITIES

In this section it is shown that the results of Oi and of Tisdell can be generalised to cover the case when the prices of a number of commodities, interdependent in production are subject to variation. To do this let X represent the vector of the quantities of commodities (products plus inputs) involved in a perfectly competitive firm's production and let P represent the vector of their corresponding prices. Using the Hicksian convention of treating inputs as negative commodities, the firm's profit in any single period of time is

$$\Pi = PX. \quad \text{(AI.13)}$$

Assume that the prices of all the commodities are stable at values indicated by the vector, $\bar{P}$. The firm then maximises

$$\Pi = \bar{P}X \quad \text{(AI.14)}$$

subject to its production possibility set. Imagine that this maximum occurs for $\bar{X}$. Then under stationary conditions, the firm's profit per period (and on average) is

$$\Pi = \bar{P}\bar{X}. \quad \text{(AI.15)}$$

---

[2] Note that fixed costs have been ignored throughout. This does not affect the result.

[3] This result holds provided that the value of the marginal product of the variable input eventually declines [4].

In particular note that the stationary conditions imply that the production set does not alter with the passage of time.

Suppose that the prices of one or more commodities become unstable but maintain the same value on average so that the vector of average commodity-prices remains unchanged at $\bar{P}$. Let $\bar{R}$ represent the vector of relative prices implied by $\bar{P}$ when the price of one of the commodities is taken as a *numeraire* and let R represent the vector of relative prices implied by P when the same commodity is used as a *numeraire*. Then if in any period $R \neq \bar{R}$, the average profit of the firm must as a rule increase, and cannot decrease.

The average profit of a firm cannot decrease because it has always the option of adopting production strategy $\bar{X}$ in each period and, on average, this ensures the firm of a profit of $\bar{\Pi}$. However, if in any period $R \neq \bar{R}$, the firm will as a rule be able to make more profit by diverging from $\bar{X}$ than by maintaining this configuration of production. Consequently, as a rule the firm will benefit from (relative) price instability of factor and/or product prices and its benefit will be greater as a rule the greater is the range of commodities subject to price instability.

The condition for the firm to gain from price instability is that there be at least one period in which $\bar{X}$ fails to maximise the firm's profit. If the firm diverges from $\bar{X}$ on this occasion this must increase its profit and raise profit on average above $\bar{\Pi}$. The greater the scope for profitable divergence from $\bar{X}$, either prices diverge from $\bar{P}$ more frequently or because more commodities become subject to price instability, the greater is likely to be the addition of instability to profit on average.

Under Hicksian type of production conditions in which any changes in relative prices alter the configuration of production, average profit is increased by price instability of inputs and/or products. However, when $\bar{X}$ is situated at a corner-point of the production set (when substitutability in production is impeded, as may be indicated for instance by inelasticity in the supply curve), average profit may fail to rise as a result of price instability. But in the normal case, where some substitution is possible, the average profit of firms can be expected to rise as a result of price instability provided that *relative* prices are subject to variation. If prices rise and fall in the same ratio, average profits are, of course, not affected by price instability because $\bar{X}$ remains optimal, and average prices are unaltered.

It is not surprising that price instability in a world of certainty and a stationary production set has the results outlined above. Under price instability the firm can always do as well on average as under price stability. However, by taking full advantage of price variations and adjusting its production, a firm can usually do better on average under price instability.

The theorem is quite general and the more factors and the more products subject to price variability the greater on average is the gain of the firm.[4]

REFERENCES

[1] OI, W.Y., "The Desirability of Price Instability Under Perfect Competition", *Econometrica*, 29, 58-64, 1961.

[2] TISDELL, C.A., "Uncertainty, Instability and Expected Profit", *Econometrica*, 31, 243-247, 1963.

[3] TISDELL, C.A., *The Theory of Price Uncertainty, Production and Profit*, Princeton University Press, Princeton, 1968.

[4] TISDELL, C.A., "Price Instability and Average Profit", *Oxford Economic Papers*, 22, 1-12, 1970.

[5] TISDELL, C.A., "Influences of Variable Product-Prices on the Incomes of Multi-Product Firms and Agricultural Economics", *Journal of Agricultural Economics*, 27, 227-233, 1976.

[6] TISDELL, C.A., "Extension of Oi's True Instability Theorem", *Journal of Economic Theory*, (forthcoming).

[4] The proof of this theorem was originally indicated in [5] and extended in [6].

APPENDIX II

Does Price Instability Increase Consumers Welfare as Waugh and Massell Suggest?

AII.1 INTRODUCTION

In 1944 Waugh [10], [11], showed using a partial one-product model that price instability raises consumers surplus on average if consumers are perfectly knowledgeable about prices and, therefore, the view has gained support that price instability raises consumers welfare. Of course, this may not be so in a world in which price uncertainty is important and increases with price variability [8], nor need it hold in a world like Simon's [6], in which decision-making involves a cost which may rise as the price of a product becomes more variable. But even putting these important complications to one side, it is doubtful whether price instability can be reasonably claimed to raise consumers welfare, even though the logic of Waugh's argument concerning consumers surplus appears correct.

It is argued in this Appendix that under the conditions envisaged by Waugh, price instability may reduce the average level of utility obtained by consumers and that there is always a set of stable prices, when preferences are stationary and storage costless, which are superior to any unstable set. In general, under stationary conditions, price instability appears to reduce consumers welfare. But before developing this thesis, it is useful to restate Waugh's proposition.

AII.2 WAUGH'S ARGUMENT RESTATED

Waugh argues that if the variability of the price of a product about a constant average value rises, other prices unchanged, consumers surplus per period rises on average if consumers' tastes are constant and if consumers are perfectly knowledgeable about prices. This argument can be illustrated by means of Fig. AII.1.

At a stable price of E[p] for the product X, consumers surplus per period is equivalent to the area of triangle B plus the area of the hatched quadrilateral. If the price of the product begins to fluctuate and is p_1 half of the time and p_2 in the remainder so that the product's average value remains unchanged at E[p], consumers surplus when price is p_2 is equal to the area of triangle B and when price is p_1 it is equal to the area of triangle B plus areas of the hatched and the shaded quadrilaterals plus the area of triangle G. Thus, for these price fluctuations consumers surplus per period is *on average* equal to the area of triangle B plus the area of the hatched quadrilateral plus half of the area of triangle G. Consequently, consumers surplus per period is on average higher by the area of half triangle G in the case where prices fluctuate than in the case where they are stable.

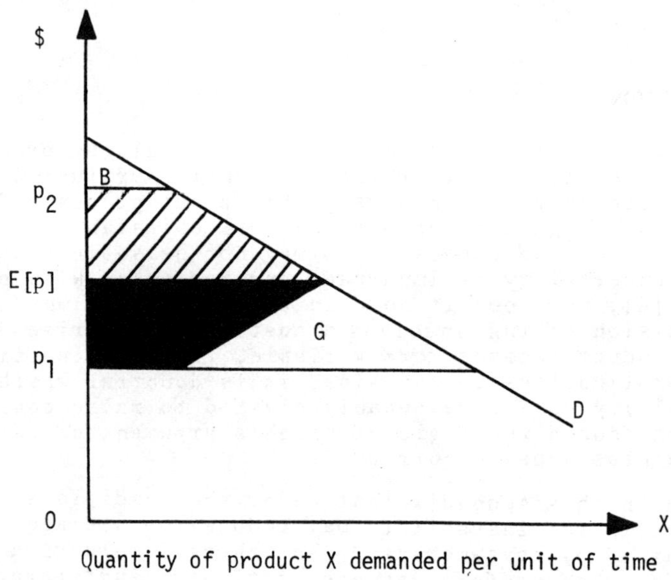

Fig. AII.1

The above result generalises easily when the demand curve for a product is linear. Letting X represent the quantity demanded of product X and letting p indicate its price, consider the consequences of a demand curve of the form

$$X = \alpha - \beta p. \qquad \text{(AII.1)}$$

For such a demand curve, consumers surplus for any price of X is represented by the area of an appropriate triangle, an example of which is illustrated by the hatched area in Fig. AII.2. Consumers surplus, C.S., represented by the area of this triangle is

$$C.S. = 1/2(\alpha/\beta - p)X \qquad \text{(AII.2)}$$

$$= 1/2(\alpha/\beta - p)(\alpha - \beta p) \qquad \text{(AII.3)}$$

$$= 1/2(\alpha^2/\beta - 2\alpha p + \beta p^2) \quad \text{(AII.4)}$$

a quadratic function in p. Since $E[p^2] = E[p]^2 + \text{var } p$, consumers surplus on average can be expressed if p is subject to variation, as

$$E[C.S.] = 1/2(\alpha^2/\beta - 2\alpha E[p] + E[p]^2 + \beta \text{ var } p). \quad \text{(AII.5)}$$

Therefore, if the average price of the product is constant, the average level of consumers surplus rises as the variance in the price of the product increases because

$$\frac{\partial E[C.S.]}{\partial \text{ var } p} = 1/2\beta > 0. \quad \text{(AII.6)}$$

The flatter the demand curve is for a product the greater is the increase in the average level of consumers surplus resulting from an increase in the variability of the product's price.

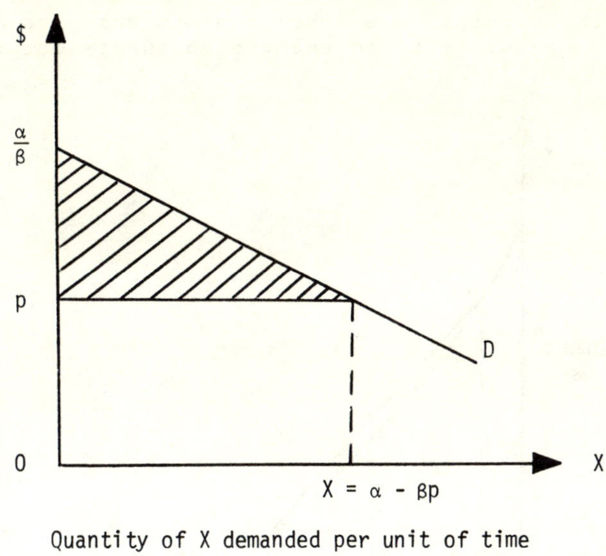

Fig. AII.2

Using the properties of convex functions, Waugh's proposition can be generalised to cover any normally sloped demand curve. If the demand curve for product X is negatively sloped, consumers surplus as a function of the price of X falls at a decreasing rate with increases in p. Expressed as a function of the price of X, consumers surplus is

$$C.S. = F(p), \quad F'(p) < 0, \quad F''(p) > 0 \quad \text{(AII.7)}$$

so that consumers surplus is strictly convex function of p. From the properties of strictly convex function [1, p. 74],

$$E[F(p)] > F(E[p]) \quad \text{(AII.8)}$$

if not all values of p = E(p). If the price of X fluctuates consumers surplus on average exceeds that when price is stable at the average of its fluctuating values.

The nature of this result can be illustrated by means of Fig. AII.3. The curve BAD represents consumers surplus as a function of the price of X. Suppose that the price of product X is p_1 half of the time and p_2 in the remainder. Consumers surplus when the price of the product is p_1 corresponds to B and when the price of the product is p_2 consumers surplus corresponds to D so that on average consumers surplus corresponds to C, the mid-point of the chord linking B and D. However, if the price of product X is stabilised at E[p], the average value of p_1 and p_2, consumers surplus per period falls to a level corresponding to A. Stabilisation of the product's price reduces consumers surplus on average.

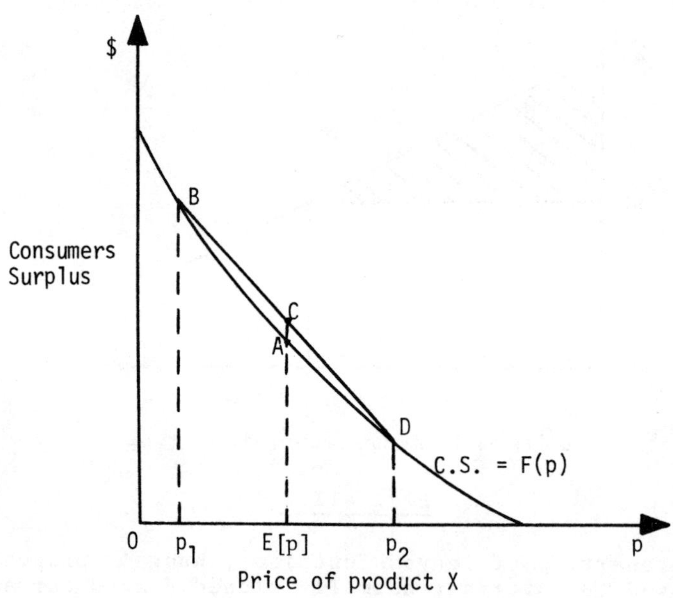

Fig. AII.3

Whether or not the compensating variations or equivalent variation measures of consumers surplus based upon a compensated demand curve are used or Marshallian consumers surplus based on a conventional Marshallian demand curve, uncompensated for income effects, is used, Waugh's result still follows provided that the demand curve is downward sloping. However, it is doubtful whether consumers surplus is a satisfactory measure of consumers welfare [5], and whether price instability raises consumers welfare.[1] This can be initially illustrated by a simple example.

AII.3 ONE PRODUCT ILLUSTRATION OF A REDUCTION IN CONSUMERS WELFARE AS A RESULT OF PRICE INSTABILITY

Imagine that a consumer purchases only one product and spends all of his income on the purchase of the product. His demand curve for the product therefore forms a rectangular hyperbola like that shown in Fig. AII.4. The curve is possibly non-existent below a minimum quantity of product X, $\hat{x}$ because a minimum amount of the product is required for the consumer's survival. In this case, the full impact of price variations fall on the consumer's income or his consumption of product X.

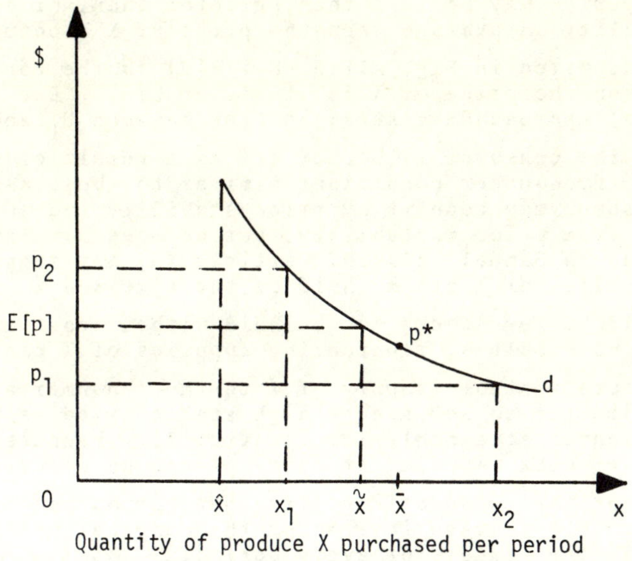

Fig. AII.4

Given the above circumstances, we shall show that (i) under conditions similar to those assumed by Waugh, price instability *may* reduce the average utility obtained by a consumer from his income

[1] The following results were discovered independently of Samuelson's 1972 contribution [5] and our presentation differs from his.

and (ii) if storage of the produce is costless, there is a possible stable price for the produce which is preferred by the consumer to the set of fluctuating prices. The latter result in particular throws doubt on the value of Waugh's conclusion.

The argument can be illustrated by reference to Figs AII.4 and AII.5. The curve marked $U = f(x)$ in Fig. AII.5 represents a consumer's von Neumann and Morgenstern utility curve. When the price of X is stable at $E[p]$ the consumer purchases $\tilde{x}$ of X per period and his level of utility per period is $\tilde{U}$. If the price of X is p_1 fifty per cent of the time and p_2 in the remainder, his level of utility when p_1 is the case is U_1 and when p_2 is the case it is U_2 so that on average his utility is $E[U]$. Since the consumer's demand curve declines at decreasing rate the consumer's consumption of X per period when price is stable at $E[p]$, $\tilde{x}$, is less than his level of consumption on average, $\bar{x}$, when price is unstable. Depending upon the shape of $U = f(x)$, $\tilde{U}$, the consumer's utility per period may be less than, greater than, or equal to $E[U]$, his utility on average when the price of X fluctuates. In the case illustrated in Fig. AII.5, $\tilde{U} > E[U]$ so the consumer is better off when the price of X is stable at $E[p]$. But if the curve $U = f(x)$ approaches a straight line between U_1 and $\bar{U}$, $\tilde{U} < E[U]$ and the consumer is better off as a result of price instability. Even under conditions similar to those assumed by Waugh, a consumer may benefit by price stability and suffer a welfare loss from price instability, but he does not invariably do so even though Samuelson's 1972 article [5] may suggest otherwise. If supplies of X are x_1 half of the time and x_2 in the other half, if buffer stocks can be held without cost and carry-back creates no problems, fluctuating supplies of X can be used always to ensure a stable supply of $\tilde{x}$ to the consumer at a stable price of $E[p]$. But to set such a high stable price is to forgo needlessly greater attainable levels of utility because a stable supply of $\bar{x} = 0.5x_1 + 0.5x_2$ at price p^* can be assured. At the *stable price* p^* the consumer's utility per period reaches its highest attainable average level[2] for the marginal utility from consuming X is then equal for every period.[3] Utility per period is $\bar{U}$.

---

[2] If the utility function is strictly convex and stationary and a given supply or stock is to be allocated for an interval of time, the supply must be equally allocated to each period to maximise total utility for the interval.

[3] Note that in this case $p^* < E[p]$, Waugh's average price. This inequality holds if the demand curve is strictly convex. If the demand curve is linear $p^* = E[p]$. In either case stabilisation at $E[p]$ may be possible. But if the demand curve is concave, stabilisation at $E[p]$ would result in a demand for more than the average level of available supplies.

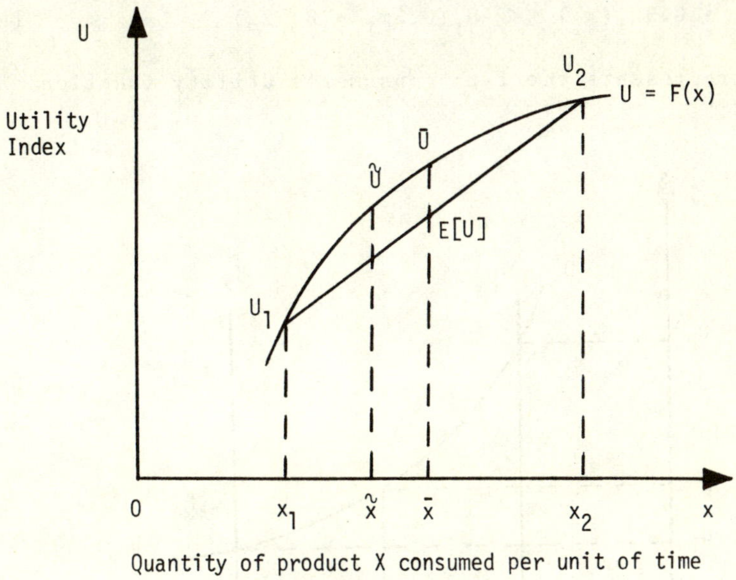

Fig. AII.5

The above discussion can be extended to show that Massell's conclusion [4] that consumers lose from price stabilisation if the source of price instability is shifts in the supply curve, is misleading. Massell assumes a system of self-liquidating buffer stocks, that the holding of such stocks is costless and that carryback creates no problem. As in Waugh's case, Massell's argument depends on consumers surplus as measured by the area under a demand curve.

Consider the case illustrated by Fig. AII.6. In this example all consumers are assumed to have given incomes, to spend their incomes all on X and to experience diminishing marginal utility from its consumption. The demand curve for product X is, therefore, represented by the rectangular hyperbola DD in Fig. AII.6. Furthermore, suppose that the supply curve is S_1 half of the time, and S_2 in the remainder so that in the absence of storage the price of X is p_1 half of the time and p_2 in the remainder. But a self-liquidating buffer stock system is feasible and enables consumers to be charged a stable price of p* for X. Producers may also be paid this price of X, or can be paid p_2 when S_1 is the case and p_1 when S_2 is the case. The stabilised price clearly raises the average level of utility experienced by consumers for their utility functions are strictly concave functions of X. For the i-th consumer and in view of the strict concavity of utility functions

$$0.5U_i(x_1) + 0.5U_i(x_2) < U_i(0.5x_1 + 0.5x_2) \qquad \text{(AII.9)}$$

where U_i represents the i-th consumer's utility function.

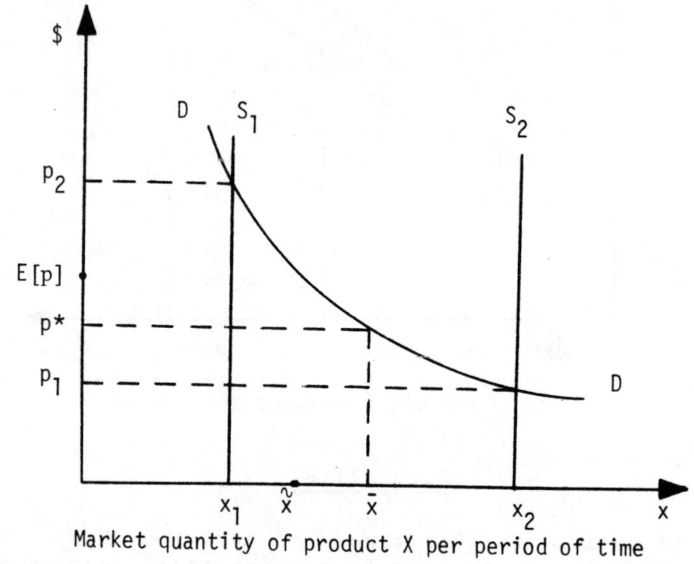

Fig. AII.6

In view of the above findings, the conclusion of Massell that consumers lose from price stabilisation, if the source of price instability is supply shifts, is completely misleading.[4] The theorems of Waugh and Massell are misleading basically because they rely on consumers surplus to measure consumers welfare - an error that could lead to fallacious policy consequences.[5]

---

[4] Note that the above result applies to any additive disturbances of the supply curves. The supply curves S_1 and S_2 need not be vertical for the above result to hold. It is sufficient for these curves to be parallel to one another.

[5] It is unfortunate that some writers continue to develop the theory of welfare gains from price stabilisation using expected consumers surplus as a measure of consumers welfare gains. Turnovsky, for example, does mention in a footnote [9, p. 134] that the use of consumers surplus to measure consumers welfare has various shortcomings, but he develops his model regardless of this observation.

AII.4 CONSUMERS GAINS FROM PRICE STABILITY IN A MULTIPRODUCT CONTEXT

The previous argument applies *mutatis mutandis* to the purchase of several products by consumers if the prices of all products rise and fall together in the same proportion and if the income-expansion paths of consumers are linear, as they will be if consumers' utility functions are homogeneous. This can be illustrated by means of the example shown in Fig. AII.7.

Suppose that a consumer consumes two products x_1 and x_2 and that the *non-controlled* supplies of these commodities and their consequent prices are such that the consumer's budget line is AM_1 half of the time and CM_3 half of the time, given his constant money income. The supplies to the consumer therefore vary between A and C. But this is not the attainable optimum from the consumer's viewpoint, if storage is costless and carryback creates no problem.

In this case, it is optimal to supply the consumer with quantity B of the products in each period at *stable* prices corresponding to budget line BM_2 and to use buffer stocks to make this possible. A buffer stock scheme (in which producers are paid these stable prices for their supplies) is self-liquidating because AB = BC. The stable solution is optimal from the consumer's point of view if his utility function is strictly concave for then the segment of it corresponding to the line AC is strictly concave and consequently

$$0.5U(A) + 0.5U(C) < U(B) \tag{AII.10}$$

and in addition, maximum utility is only achieved by equally distributing available supplies between periods.

However, this theory can be generalised further. Assume that during an interval of time of n periods that supplies of the r products (of some at least) are subject to *non-controlled* movements only. Let the vector of these supplies in period t be represented by X_t.

Suppose that each consumer has a Hicksian preference system (strictly concave utility function at least twice differentiable, all products purchased and so on),[6] that his tastes are unchanged during the time interval and his money income is constant. Then if the prices of any of the products show any variation during the time interval, the budget line of each consumer is not the same throughout the time interval. Consequently his consumption bundle is not the same throughout the time interval. But if this is so, and the carrying of stocks is costless and if carryback

---

[6] These conditions can be relaxed to some extent.

creates no problem, utility on average is lower than is attainable. If consumption bundles differ in the periods for any consumer it follows from the properties of strictly concave functions that greater average (expected) utility can be obtained by ensuring a more even (on average) flow of supplies. Given the costless properties of stocks, no consumer obtains maximum utility unless his consumption bundle is the same in every period. To ensure this his budget line must be the same in every period and prices must be stable to ensure this.

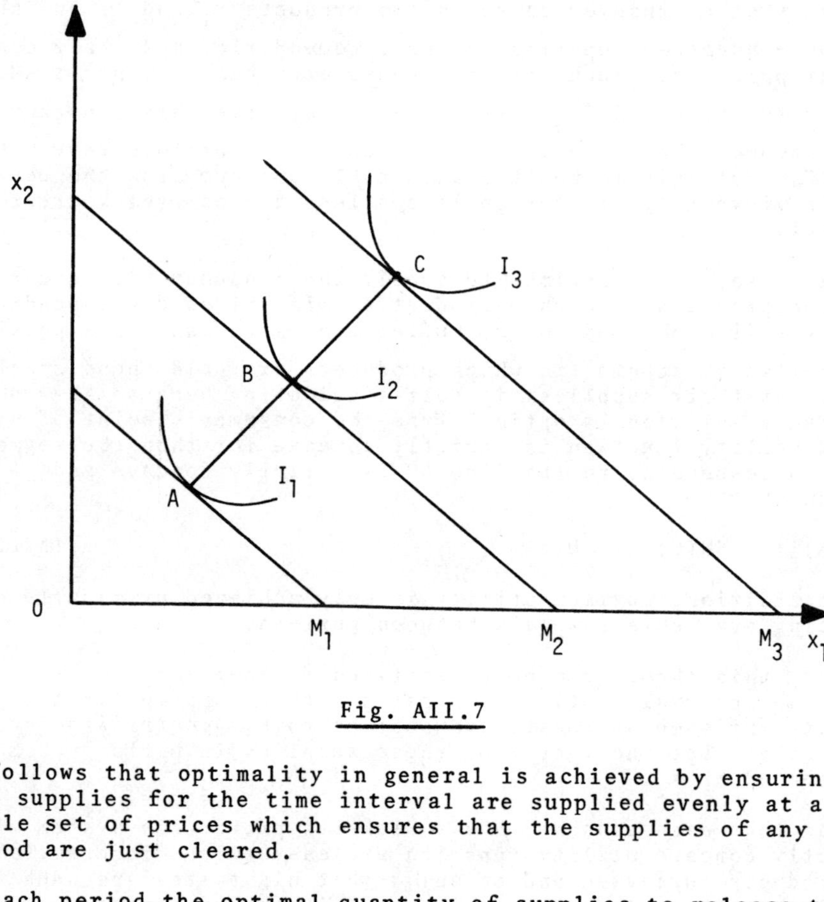

Fig. AII.7

It follows that optimality in general is achieved by ensuring that supplies for the time interval are supplied evenly at a stable set of prices which ensures that the supplies of any period are just cleared.

In each period the optimal quantity of supplies to release to buyers is

$$\frac{1}{n}\sum_{t=1}^{n} X_t$$

and the optimal set of stable prices P_t^* is the set which just clears these supplies.

The above conclusion is quite general. It allows for the possibility of variations in the price of only one-product or in several. It might be observed also that the optimal set of stable prices is not usually the average of the fluctuating ones which are an alternative.

Given the non-controlled nature of supplies, price stability is achieved in the above case through a self-liquidating buffer scheme. Producers can be paid P_t^* in each period for their supplies, which are variable. The payment to producers for the time interval is

$$\sum_{t=1}^{n} P_t^* X_t$$

which equals the total receipts from consumers.

If stocks do involve a cost or if carryover is a problem, absolute price stability is no longer desirable as a rule although it still may be desirable to use a buffer stock scheme to moderate price fluctuations. Furthermore, if production can be controlled this needs to be taken into account. Nevertheless, if stocks involve no cost, carryback is no problem and production is subject to *some* non-controlled variation, then in stationary conditions, a stable set of prices is optimal from the viewpoint of consumers and can be achieved through self-liquidating buffer stocks. Otherwise, Pareto optimality is not achieved and consumers are not as well-off as is possible. Taking account of all market participants (producers and consumers) price stability is necessary for Pareto optimality in the above circumstances.

While Oi's proposition that price instability increases the average profit of perfectly competitive firms, provided that the average price of the product is constant, is correct and generalises to the n-commodity case, Waugh's corresponding view that price instability raises consumers welfare does not appear to hold. Furthermore, Massell's hypothesis that price stabilisation reduces consumers welfare when instability arises from non-controlled variations in supply is grossly misleading. Price stabilisation can be Pareto optimal when supplies are subject to variation and perfect knowledge exists.

Therefore, one must agree with Samuelson that:

> Price stability is, other things equal, in itself a definite virtue. This is so whether one takes the view of producers or consumers or both, in the following sense: Under stable conditions of supply and demand, or more basically of technology and tastes, departures from price stability usually violate Pareto optimality, to the extent that producers or consumers gain, the loss to other parties is 'greater' than their gain. [5, p. 493]

The case for feasible price stability may be strengthened even further if price instability generates price uncertainty and adds to the cost of decision-making.

REFERENCES

[1] HARDY, G.H., LITTLEWOOD, J.E., and POLYA, G., *Inequalities*, Cambridge University Press, Cambridge, 1934.

[2] OI, W.Y., "The Desirability of Price Instability under Perfect Competition", *Econometrica*, 29, 58-64, 1961.

[3] OI, W.Y., "The Consumer Does Benefit from Feasible Price Stability: A Comment", *Quarterly Journal of Economics*, 86, 494-498, 1972.

[4] MASSELL, B.F., "Price Stabilisation and Welfare", *Quarterly Journal of Economics*, 83, 284-298, 1969.

[5] SAMUELSON, P.A., "The Consumer Does Benefit from Feasible Price Stability", *Quarterly Journal of Economics*, 86, 476-493, 1972.

[6] SIMON, H., *Administrative Behaviour*, Macmillan, New York, 1961.

[7] TISDELL, C.A., "Uncertainty, Instability, Expected Profit", *Econometrica*, 31, 243-247, 1963.

[8] TISDELL, C.A., *The Theory of Price Uncertainty, Production and Profit*, Princeton University Press, Princeton, 1968.

[9] TURNOVSKY, S.J., "The Distribution of Welfare Gains from Price Stabilization: A Case of Multiplicative Disturbances", *International Economic Review*, 17, 133-148, 1976.

[10] WAUGH, F.V., "Does the Consumer Benefit from Price Instability", *Quarterly Journal of Economics*, 57, 602-614, 1944.

[11] WAUGH, F.V., "Consumer Aspects of Price Instability", *Econometrica*, 34, 504-508, 1966.

Author Index

Subject Index